Met. O. 887

METEOROLOGICAL OFFICE

THE
MARINE OBSERVER'S
HANDBOOK

10TH EDITION

LONDON
HER MAJESTY'S STATIONERY OFFICE

551.501.1 (02)

HER MAJESTY'S STATIONERY OFFICE

Government Bookshops

49 High Holborn, London WC1V 6HB
13a Castle Street, Edinburgh EH2 3AR
41 The Hayes, Cardiff CF1 1JW
Brazennose Street, Manchester M60 8AS
Southey House, Wine Street, Bristol BS1 2BQ
258 Broad Street, Birmingham B1 2HE
80 Chichester Street, Belfast BT1 4JY

Government publications are also available
through booksellers

£5·50 net

Printed in Scotland by Her Majesty's Stationery Office at HMSO Press, Edinburgh
Dd 497088 K36 3/77 (13855)

ISBN 0 11 400297 5

CONTENTS

iii

CONTENTS—*continued*

TABLES

iv

ILLUSTRATIONS

FOREWORD

The Marine Observer's Handbook is written for three purposes:

(1) To assist officers aboard British vessels, who voluntarily make observations at sea on behalf of Meteorological Services, to carry out this work in an efficient and uniform manner.

(2) To encourage all mariners to take an interest in meteorology and to assist them in their study of this important and interesting subject.

(3) To provide a book of reference for candidates for Department of Trade Mate's and Second Mate's examinations.

It will be noted that this book deals with meteorological instruments and the practical aspect of making observations. The companion volume, entitled *Meteorology for Mariners*, embraces the theory and the application of meteorology to the seaman's profession.

The seaman is so dependent on the weather that an interest in meteorology on his part is essential for the safe and economic operation of his ship. It is undoubtedly true that in this modern age of large, fast, power-driven ships, just as in the days of sailing ships, no ship's officer can consider himself a complete mariner unless he is 'weather wise'. Meteorological observing tends to quicken the eye of the observer, making him more alert and more ready for emergencies. The practised observer is not only on the look-out for changes in weather and cloud and for interesting phenomena, but by his general alertness he will ensure that there are no 'Irish pennants', loose tarpaulins etc., when he is on deck.

Essentials to efficiency in meteorological observations are accuracy and attention to detail. The results are beneficial not only to Meteorological Services, and thence to mankind, but to the ship herself. By accurate reading and intelligent interpretation of humidity observations, for example, the master can decide whether ventilation of cargo is wise or not, or by a combination of sea and air temperature and humidity, the likelihood of fog can be forecast aboard the ship. Timely notice of a shift of the wind or variation of its force or the sky becoming overcast or gradual deterioration of visibility may, on occasions, save a ship from getting into difficulties. The largest and most powerful ship can be delayed or damaged by rough seas or slowed down in fog. Valuable cargoes can be quickly ruined if due regard is not paid to unfavourable weather.

Anything that is worth doing is worth doing well and this is particularly so with regard to meteorological observations. A lone ship's observation from 'somewhere in the ocean' may hold the key to an otherwise obscure meteorological situation, but it is better to have no observation than an inaccurate or erroneous one. An inaccurate observation may mislead the forecaster and, directly or indirectly as a result of that inaccuracy, an incorrect forecast can cause a small ship or aircraft to set out when a slight change of plan might lead to a smoother and safer passage.

Accuracy is just as important for climatological purposes. In the analysis of meteorological records for the compilation of atlases and for scientific investigation generally, a few inaccurate observations may so bias the results as to tend to falsify the picture. In deciding whether to reject an apparently erroneous observation, the investigator can only use his judgement and experience.

Observers at sea would perhaps be surprised at the many uses to which their observations are put, both commercially and scientifically. To mention only a few: frequencies of winds of gale force are required whenever the load line areas are reviewed; air temperatures and humidities have been useful in the testing of life jackets; meteorological data are needed by the respective research organizations in connection with the design of ships and with the efficiency of radar.

Thus, by taking an intelligent interest in meteorological observations, the seaman contributes to the cause of science and benefits the world in general, and his fellow seamen in particular, by increasing our knowledge of meteorology and climatology.

Note to the tenth edition. This edition is principally an up-dated version of the ninth edition. The metric system has been introduced throughout although nautical miles and knots have been retained to conform to navigational requirements. References to cloud heights are still made in feet in accordance with the World Meteorological Organization regulations and the United Kingdom Meteorological Office practice. Chapter 2 has been extended to include distant-reading equipment and electrical resistance thermometers. The part of Chapter 7 headed Ocean Current Observations has been entirely rewritten to remove anomalies, and the Ice section has been amended to use the latest World Meteorological Organization nomenclature. Chapters 10, 11 and 12 on Phenomena have been revised and extended to include the latest information. We are indebted to the Director of the Appleton Laboratory at Slough and Dr. P. Herring of the Institute of Oceanographic Sciences at Wormley for advice and additional text.

MARINE DIVISION,
  METEOROLOGICAL OFFICE.
July 1976.

Part I   Instrumental Observations

CHAPTER 1

Atmospheric Pressure

The instruments most commonly used for measuring the pressure of the atmosphere are the mercury and aneroid barometers. Until recently, the consistent accuracy required for scientific purposes and for official meteorological work could be achieved only by use of the mercury barometer because of certain inherent errors in the aneroid instrument. The marine-type mercury barometer has been in regular use aboard British observing ships since 1854 and has been so successful that its design has been virtually unchanged through the years.

Precision aneroids have now been adopted as the standard Meteorological Office atmospheric pressure reading instrument for issue to voluntary observing ships as they proved to be of similar accuracy yet more compact in size and easier to read than the marine mercury barometer which was subject to 'pumping'. The change-over to this instrument is not yet complete and a number of selected ships may still carry mercury barometers; these will eventually be replaced. Both instruments are described below.

The mercury barometer. The principle of the mercury barometer was discovered by Evangelista Torricelli in 1643.

A simple mercury barometer (see Figure 1) is made by completely filling with mercury a glass tube closed at one end and approximately 1 metre in length. The open end is then immersed in a cistern also containing mercury, and the tube is held upright. The mercury column falls, leaving a vacuum at the top of the tube.

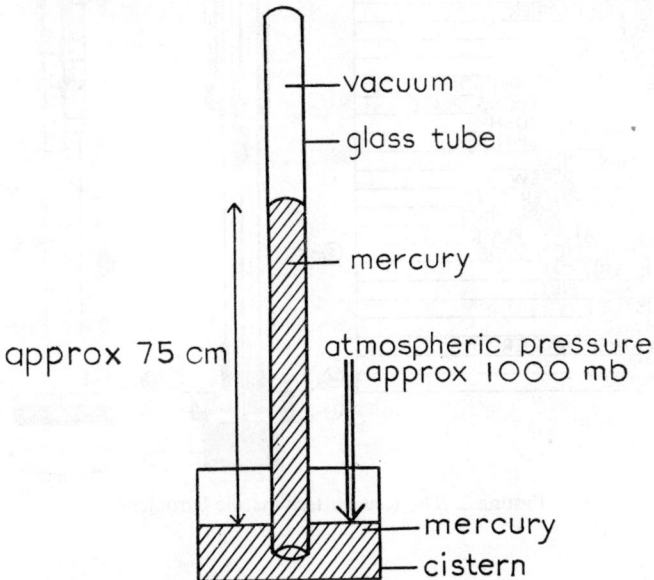

FIGURE 1. A simple mercury barometer

1

1	TOP CAP
2	SCREWED CORE
3	LEATHER WASHER
4	GLASS COVER
5	BARREL
6	VERNIER
7	MARINE TUBE
8	VERNIER CARRIER
9	RACK
10	MERCURY
11	BOTTOM CAP
12	PINION GEAR OPERATING KNOB
13	" " FIXING PLATE
14	" " BUSH
15	" " SPINDLE
16	GIMBAL RING
17	" SCREW
18	" OUTER RING
19	" LOOSE RING
20	TEMPERATURE PLATE
21	CORRECTION SLIDE
22	CISTERN CAP
23	" NUT
24	CISTERN
25	GLAND
26	LEATHER WASHER
27	ARM
28	BRACKET

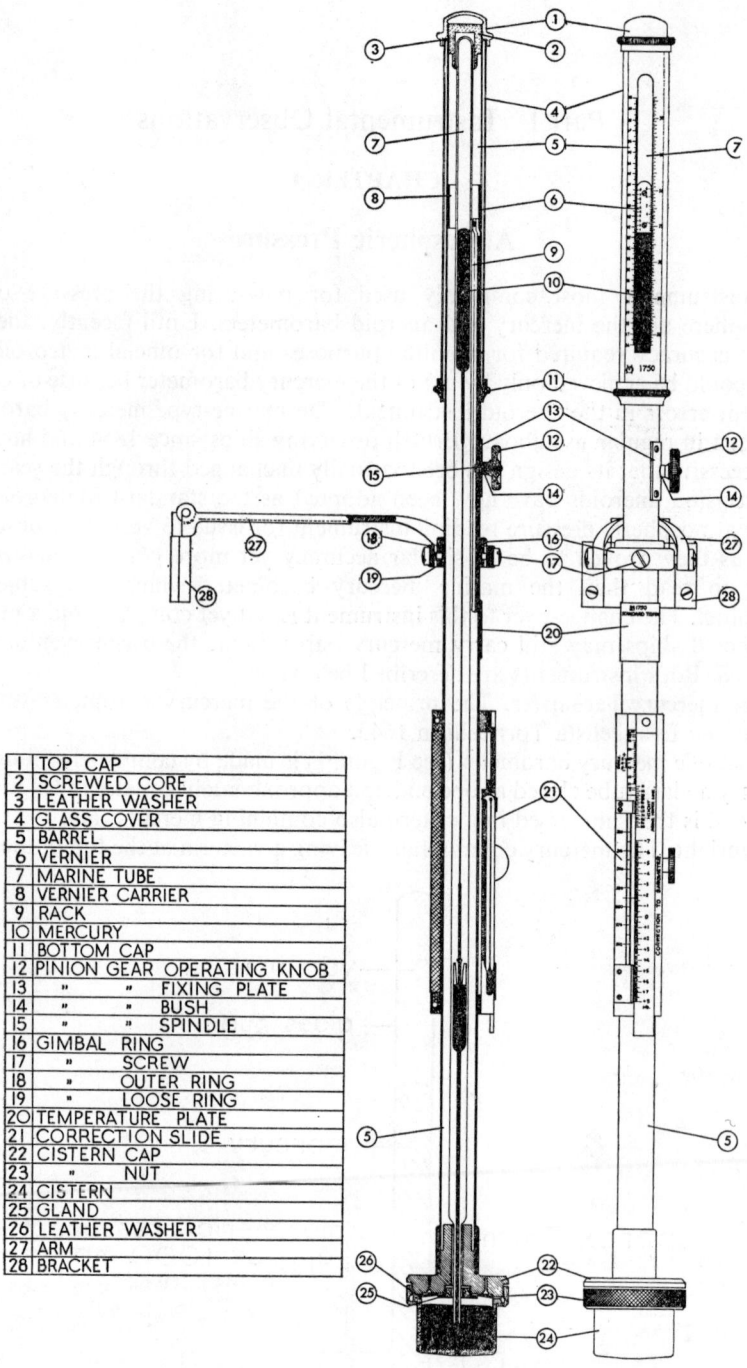

FIGURE 2. The Kew-pattern marine barometer

2

until the weight of the mercury column *above the level of the mercury in the cistern* just balances the atmospheric pressure which is exerted on the free surface of the mercury in the cistern.

The mercury barometer only gradually passed from this original simple form to that of a practical and portable instrument and was not used by seamen until a century had elapsed.

The Kew-pattern marine barometer. This consists of a glass tube and cistern enclosed in a metal protecting case (see Figure 2). In the upper part of the cistern are one or more small holes which admit the air, and a leather washer, permeable to air, which prevents the mercury from escaping, and also keeps out dust. The bore of the glass tube is considerably constricted for the greater part of its length, and, for part of this constriction, is reduced to a fine capillary. The object of these constrictions is to reduce the amount of 'pumping', i.e. oscillations of the top of the mercury column, caused by the movements of the ship and by gusts of wind. At the top of the mercury column the bore of the tube is greater; this minimizes the effect of 'capillarity'* on the height of the centre of the mercury column, but leaves the upper surface of the column sufficiently convex to facilitate accurate reading. An air-trap in the tube prevents air from rising into the space above the mercury column, which should be an almost perfect vacuum. On the metal protecting case is a scale, with a vernier for reading the height of the mercury.

For the purpose of ascertaining the temperature of the barometer itself, a thermometer is attached. On barometers graduated in millibars, the thermometer is graduated in degrees Absolute on older instruments, but in degrees Celsius on those made after 1 January 1955; on inch barometers, it is usually graduated in degrees Fahrenheit.

Graduation of barometer scales. From the invention of the barometer until comparatively recent years the reading was invariably expressed as *the height of the mercury column necessary to balance the atmospheric pressure at that instant.* In the British Isles, atmospheric pressure was therefore expressed in inches and decimals of an inch, while countries using the metre as a unit of length gave the pressure in millimetres and decimals of a millimetre. The graduations are marked on a metal scale at the side of the instrument. Barometer scales graduated in inches are readable by vernier to a thousandth of an inch (0·001).

When we express the pressure in terms of inches or millimetres of mercury, we mean that the pressure is equal to the weight per unit area of a column of mercury of that height. If water were used as the liquid the barometer would have to be approximately 10 metres long; mercury, which is used in practice on account of its high density, only needs a tube of a little less than 1 metre.

The original procedure in measuring barometric pressure, which is still often used, was to measure the height of the mercury column and to give that height, in inches or millimetres, duly corrected, as a measure of the atmospheric pressure.

---

*Capillarity is the tendency of liquids in narrow tubes to rise above or fall below the hydrostatic level. This tendency depends on the relative attraction of the molecules of the liquid for one another and for the molecules of the material of the tube. The narrower the tube, the greater the tendency to rise or fall, so that the effect is particularly well marked in hair-like or capillary tubes, hence the name 'capillarity'. If the liquid wets the solid material, it will rise in the tube, but if not, it will be depressed. In the case of water in a glass tube, therefore, the water column is raised, particularly at the edge, while the reverse is the case with mercury in a glass tube, for mercury does not wet glass.

It was later thought preferable that pressures should be expressed in units of pressure, not in units of height.

Pressure is force per unit area and the measurement of a force is the acceleration it would give to a body of unit mass which is free to move. In the International System of Units (SI)*, now adopted by most countries, the unit of force is the newton (symbol N), the force which, if applied to a mass of 1 kilogram will produce an acceleration of 1 metre per second. The unit of pressure is the pascal (symbol Pa) which is a force of 1 newton per square metre. However, the use of the millibar as a unit of pressure is so universal in the field of meteorology that it is unlikely to be superseded in the near future. A millibar is equal to 100 pascals.

$$1 \text{ mb} = 1 \text{ millibar} = 100 \text{ pascals} = 1 \text{ hPa}.$$

Barometers graduated in millibars have a longer line at each tenth millibar. By means of the vernier the pressure can be read to one-tenth of a millibar (see Figure 17). One thousand millibars equal one bar. This is equivalent to a pressure of 29·53 inches, or 750·1 millimetres of mercury at the standard value of gravity of 9·80665 m/s², and is thus very nearly equal to the average pressure of the atmosphere at sea level. An increase of one millibar (0·0295 inch) in atmospheric pressure therefore indicates an increase of about a thousandth of the previous pressure.

CORRECTIONS TO MERCURY BAROMETERS

For a given pressure, the height of the mercury column of a barometer depends upon the density (and therefore on the temperature) of the mercury and on the value of the acceleration due to gravity which is dependent on the latitude and height of the place of observation. Moreover, the indication on a barometer scale set up to measure the height of the column will depend upon the temperature of the scale itself. In order to make barometer readings comparable all over the world it is therefore necessary to specify 'standard conditions'—temperature of mercury and of scale, and acceleration due to gravity—and to compute tables whereby corrections may be made for any departure from these conditions. In addition the readings have to be reduced to a standard altitude, which for altitudes of not more than a few hundred metres is mean sea level, and reduction tables have also been computed for this purpose.

The 'standard conditions' referred to above were changed with effect from 1 January 1955, by a decision of the World Meteorological Organization at Geneva in 1953. The standard value of gravity was changed from 9·8062 m/s² (which is very nearly equal to the gravity at mean sea level in latitude 45°) to 9·80665 m/s² (which is a conventional standard). For barometers reading in inches the standard temperature was changed from 32°F for the mercury and 62°F for the scale to 0°C for the whole instrument. For Meteorological Office barometers reading in millibars it was changed from 285°A† (12°C) to 0°C, for the whole instrument.

Because of these changes and because it will be many years before all the older barometers are converted or replaced it is now necessary to have two

---

*Further information on the SI system will be found in the Appendix to the Tables, at the end of this book (page 151).

†See page 22.

4

sets of correction tables available for mercury barometers, whether they be graduated in inches or millibars, one for instruments adjusted to the old standard conditions and one of those adjusted to read correctly at gravity 9·80665 m/s² and temperature 0°C.

All mercury barometers have also to be corrected for capillarity†, which tends to depress the mercury in the tube; variation in the quantity of mercury in the cistern according to the height of the mercury column; defective vacuum, etc., but these are all made very small by suitable allowances in the process of construction and any residual errors and errors due to imperfections in construction or adjustment are included in the 'index errors' of the instrument. These index errors are determined at the National Physical Laboratory for all Meteorological Office mercury barometers and are given on a certificate supplied by that institution.

It is desirable that a mercury barometer should be checked every three months. All Port Meteorological Officers have a standard barometer which is available for such comparisons and Meteorological Office tested barometers are available for comparison in many dock offices etc., in the U.K. (see *Admiralty List of Radio Signals*, Vol. 3 or commercial nautical almanacs). The ship's barometer should be corrected for temperature, altitude and latitude (i.e. gravity) before a comparison is made.

Corrections to inch barometers

CORRECTION FOR TEMPERATURE. With increase of temperature, mercury expands, so that its weight per unit volume decreases; a column of greater height will therefore be required to balance a stated air pressure when the temperature is high than when it is low. The correction required to adjust the height to that at a standard temperature is therefore negative when the temperature is above the standard and positive when it is below. For an inch barometer adjusted to the old conventions, the standard temperature at which the mercury in the barometer would read correctly is 32°F. The metal scale on the instrument. however, reads correctly at 62°F; the reason for this is that the standard inch is legally defined as a certain height engraved on brass *at a temperature of 62°F*. Table 1 gives the temperature correction for such a barometer. The net effect of the different standard temperatures for the *brass scale* (62°F) and the *mercury* (32°F) is that the temperature at which the temperature correction in the table becomes zero is below 32°F. At any other temperature, the temperature correction is proportional to the height of the mercury column, that is, to the atmospheric pressure. Table 1 therefore gives the correction for ranges both of temperature and barometer readings. The temperature to be used with this table is that of the thermometer attached to the barometer, which, if the exposure of the barometer is correct, will give the temperature both of the mercury and of the brass scale.

For an inch barometer adjusted to the new conventions the standard temperature at which both mercury and scale would read correctly is 32°F. The appropriate corrections are given in Table 2. Here also the temperature to be used is that of the attached thermometer.

CORRECTION FOR GRAVITY. The barometer reading now has to be corrected for the variation of gravity from the standard value. Owing to the flattening of the earth at the poles, the distance of its surface from its centre of gravity is

---

†See footnote on page 3.

least in those regions, and greatest at the equator. In addition, the vertical component of the centrifugal force due to the earth's rotation is greatest at the equator and decreases to zero at the poles. The force of gravity, therefore, increases steadily from low to high latitudes. The greater the force of gravity, the greater will be the weight of a given mass of mercury and hence the smaller will be the height of the column required to counterbalance a given pressure of the air. For a barometer adjusted to the old conventions, standard gravity is 9·8062 m/s², which is very nearly the force of gravity at mean sea level in latitude 45°. The height of the mercury column is accordingly corrected to what it would be in that latitude, the correction being negative in low latitudes and positive in high latitudes.

At any given latitude differing from 45°, the correction to be applied is proportional to the mass of mercury forming the mercury column of the barometer. It is therefore proportional to the height of that column. Table 3 gives the correction at all latitudes and for two readings, 29 and 31 inches.

For a barometer adjusted to the new conventions standard gravity is 9·80665 m/s² and Table 4 gives the appropriate corrections for each degree of latitude and for the two pressures, 29 and 31 inches.

REDUCTION TO STANDARD LEVEL. When the barometer reading has been corrected for temperature, it must next be corrected for height above sea level. The pressure of the atmosphere at any level is equal to the total weight of all the air above a plate of unit area, held horizontally at that level. If such a plate were moved vertically upwards, the total weight above it must decrease by the weight of the column of air through which the plate had passed, and the pressure would fall accordingly. Thus, a barometer reading of 30 inches at sea level would fall to about 28·9 inches at 300 metres, and to about 21 inches at 3000 metres. The barometer reading must therefore be reduced to a standard level, and for altitudes of not more than a few hundred metres above sea level, the standard adopted is mean sea level. As the weight of a given volume of air decreases as the temperature rises, this correction depends not only on the height above mean sea level, but also on the temperature of that air (see Table 5). The temperature to be used for this table is that of the dry bulb in the screen, i.e. the temperature of the atmosphere at the time of measurement.

Besides these corrections there is also the *index correction* to be applied.

EXAMPLE. In latitude 51°N, the barometer reads 30·240 inches at a height of 11 metres above sea level. The attached thermometer reads 60°F, the dry bulb in the screen reads 58°F, the index correction is + 0·005, and the date of the NPL certificate is 1.6.53.

		Inches
Uncorrected reading		30·240
Index correction	+	0·005
		30·245
Temperature correction for 60°F (Table 1)	−	0·085
		30·160
Gravity correction in latitude 51°N (Table 3)	+	0·016
		30·176
Height correction for 11 metres at air temperature of 58°F (Table 5)	+	0·039
Corrected barometer reading		30·215

There are simple formulae by which approximate values of the temperature and height corrections can be obtained without tables. As these may sometimes be useful, they are shown in Table 11.

Corrections to Millibar Barometers

CORRECTIONS FOR TEMPERATURE AND GRAVITY. As previously explained, the standard conditions for millibar barometers were changed from gravity 9·8062 m/s^2 and temperature 285°A to gravity 9·80665 m/s^2 and temperature 0°C (273°A) with effect from 1 January 1955. The temperature corrections for barometers conforming to the old conventions, i.e. with NPL certificates dated 31 December 1954 or earlier, are given in Table 6 and those for new barometers, i.e. with NPL certificates dated 1 January 1955 or later, in Table 7. The corresponding gravity corrections are given in Table 8 (old barometers) and Table 9 (new barometers). These tables are all similar in form to the corresponding tables for inch barometers and require no further explanation.

REDUCTION TO STANDARD LEVEL. Corrections for reducing the barometer reading in millibars to mean sea level are given in Table 10 for various heights up to 50 metres and dry-bulb temperatures in degrees Celsius.

The index correction as given in the NPL certificate must also be applied.

EXAMPLE. In latitude 27°N, the barometer reads 1017·3 mb at a height of 16 metres above sea level. The attached thermometer reads 298°A, the dry bulb in the screen reads 25°C, the index correction of the barometer is + 0·3 mb and the date of the NPL certificate is 20.2.57.

		mb
Uncorrected reading		1017·3
Index correction	+	0·3
		1017·6
Temperature correction for 298°A (Table 7)	−	4·3
		1013·3
Gravity correction in latitude 27°N (Table 9)	−	1·6
		1011·7
Height correction for 16 metres at air temp. of 25°C (Table 10)	+	1·8
Corrected pressure at m.s.l.		1013·5

The correction slide. Meteorological Office marine barometers are fitted with a correction slide. This attachment makes the use of tables unnecessary for the reduction and correction of millibar barometer readings. (See Figure 3.)

A sliding piece, movable by rack and pinion, and mounted beside the attached thermometer, carries two scales; the lower, alongside the mercury column of the thermometer, is marked 'Correction to Barometer', and the upper, 'Height above Water Line'. Alongside the upper scale is mounted a strip of metal on which is engraved a 'Latitude Scale'. On the other side of the latitude scale there is engraved, on the upper part of the 'Attached Thermometer Scale', an 'Index Scale'. The Latitude Scale is fixed in such a position relative to the Index Scale as to allow for the index error of the instrument and should not, therefore, be moved. The whole slide is clamped to the barometer.

There are a number of older-pattern scales in use in which the Index Scale is omitted but these are adjusted for index error of the instrument before issue.

Before reading the barometer, adjust the correction slide so that the height of the barometer above the water line, on the appropriate scale, coincides with the latitude of the ship on the latitude scale. The correction to be applied to the barometer reading is then read off in line with the top of the mercury column in the thermometer.

7

CONVERSION OF INCHES TO MILLIBARS

A table for the conversion of barometer readings in inches to millibars is given as Table 14. In certain instances abroad, barometric pressures given in millimetres may be encountered. No table is given for the conversion of millimetres to millibars as the conversion may be made very simply by increasing the pressure in millimetres by one-third. For example, 750·0 mm is very nearly equal to 1000·0 mb.

In barometers graduated with both millibar and inch scales, and made before 1 January 1955, the uncorrected readings taken at the same time will not be comparable. The reason for this is that the millibar graduation is constructed to give the true atmospheric pressure at its standard temperature of about 285°A (12°C) at sea level in latitude 45°, whereas the inch scale is graduated to give true atmospheric pressure at a temperature somewhat below 0°C at sea level in latitude 45°. (See explanation on page 7.) The correction for temperature is different for each scale and it is only when both readings have been fully corrected that they will agree, on conversion.

In barometers conforming to the new conventions, however, the readings on the two scales will be directly related by 1000 mb = 29·530 inches, because the standard instrumental conditions of temperature and gravity are the same.

THE POSITION, SETTING UP, AND CARE OF THE MERCURY BAROMETER

Position of the barometer. In ships, the mercury barometer is usually situated in the chart room for practical convenience. The chart room is not always the best place for the barometer as it is often in an exposed position. Furthermore, the pumping of a marine barometer is reduced to a minimum when the instrument is near the centre of the ship but, except in small vessels, this is usually impracticable.

It is not possible to give fixed rules for the precise location of the barometer as circumstances vary in different ships. The following points, however, should be observed:

(a) It must be out of the way of unauthorized persons.

(b) It must not be exposed to the direct rays of the sun.

(c) It must not be exposed to suddenly varying conditions of temperature due to causes within the ship, such as draughts of air from boilers, engine room, etc.

(d) The lighting should not be such as to cause a glare on the glass surface of the barometer tube. The light should come from behind or from the side of the observer.

Setting up the barometer. The mercury barometer is supported so that it swings in gimbals and therefore tends to remain vertical when the ship is rolling. In order to give the instrument room to swing, it is supported by a suspension arm, hinged at one end. The hinged end can either be screwed to a bulkhead or shipped into a socket screwed to a bulkhead. The height of the suspension arm should be such that the top of the mercury at its highest probable position is just below the height of the eye of an average observer. A barometer that is too high is almost certain to cause errors of parallax. (See Figure 18.)

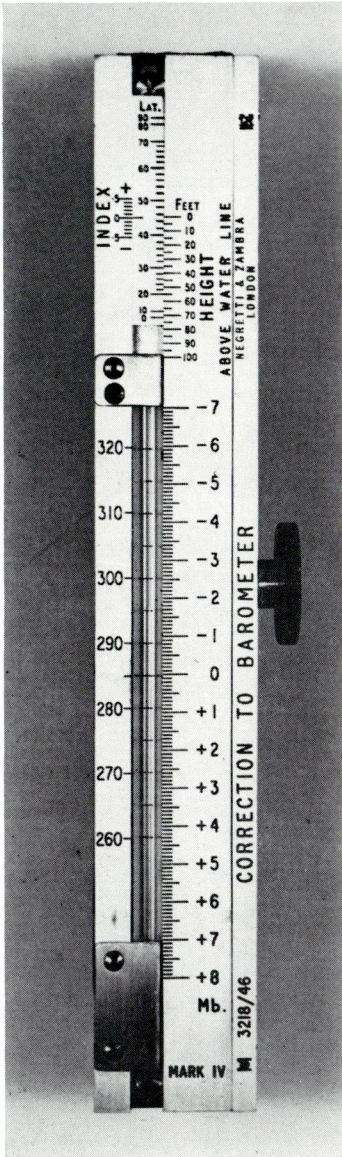

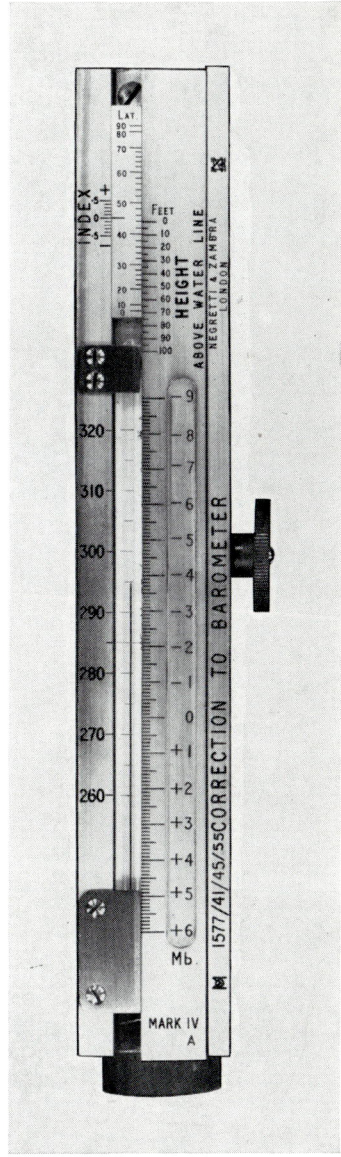

(*a*)   (*b*)

FIGURE 3. The correction slide. Both these types may still be found

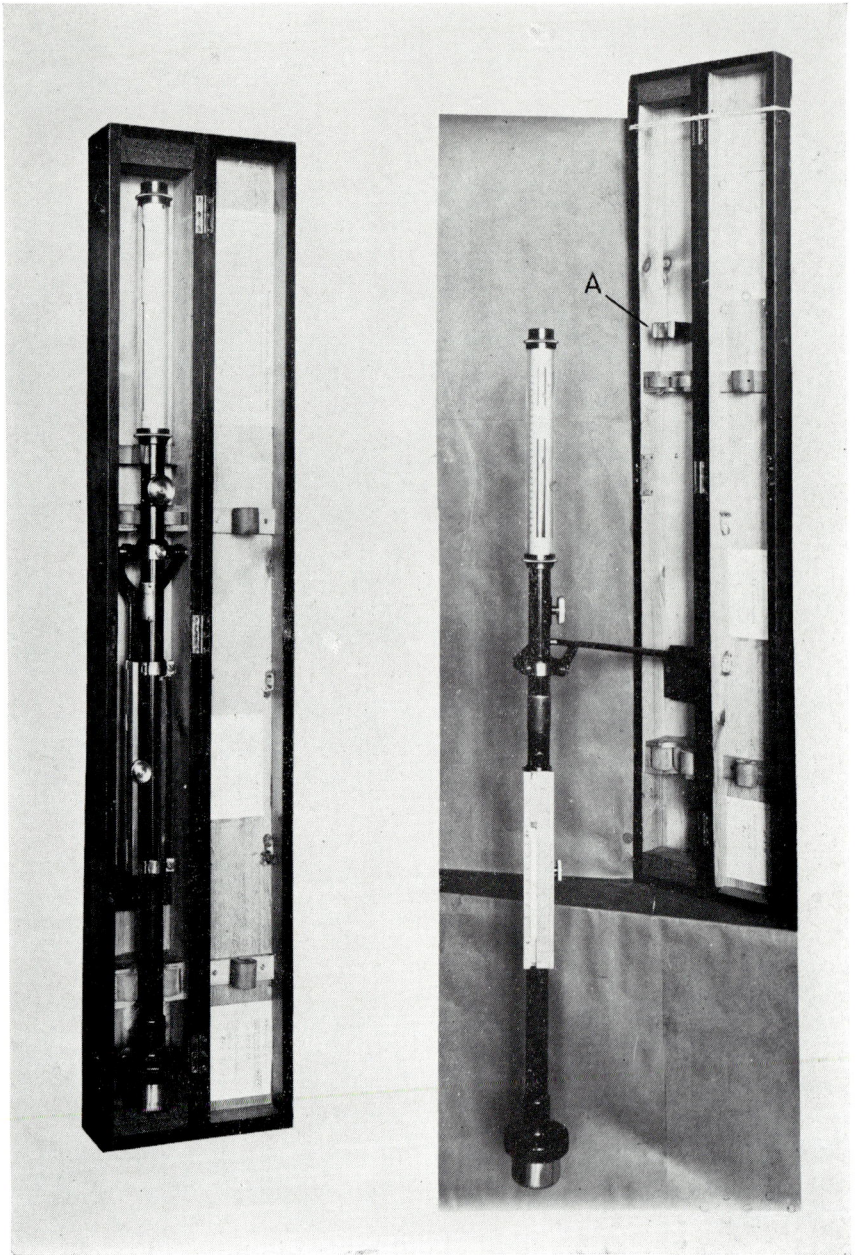

(a)   (b)

FIGURE 4. Stowage of the Kew-pattern marine barometer

FIGURE 5. The marine open-scale barograph. A description appears on pages 16–18

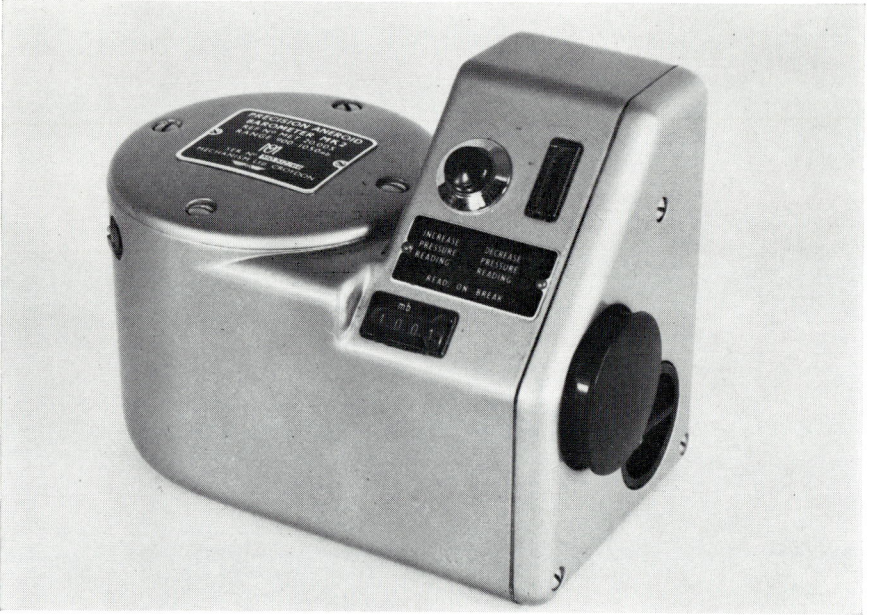

FIGURE 6(a). General view of Precision Aneroid Barometer Mk 2
Note housed static vent on left side of instrument beneath cap

FIGURE 6(b). Precision Aneroid Barometer Mk 2. View of interior

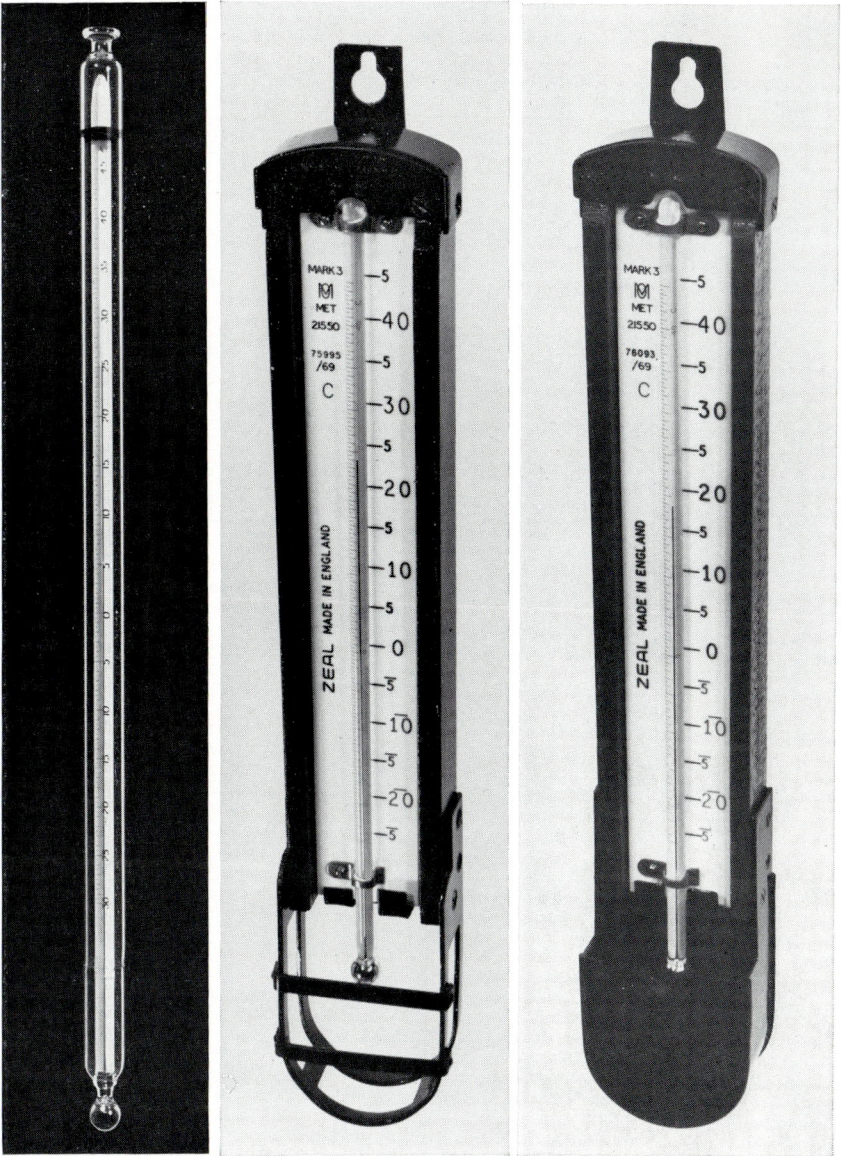

FIGURE 7. Thermometer, and air and sea protectors

FIGURE 8. The portable marine screen

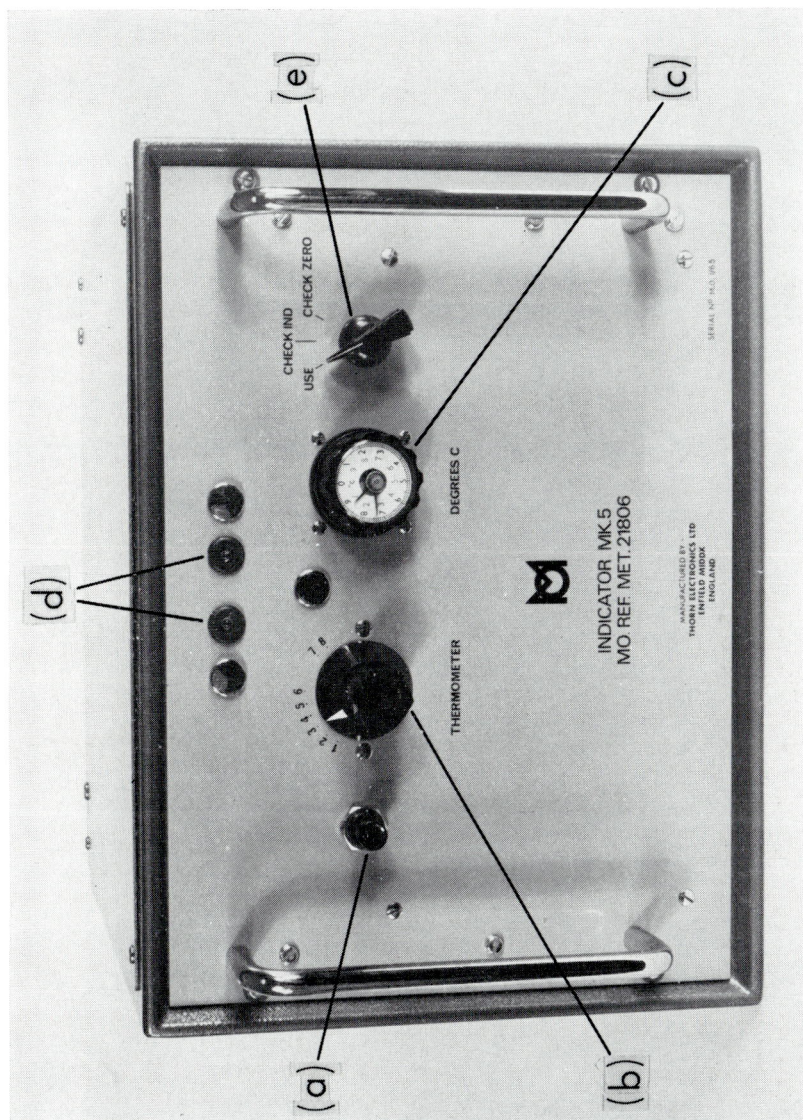

FIGURE 9. The distant-reading equipment Indicator Mk 5. A description appears on page 27

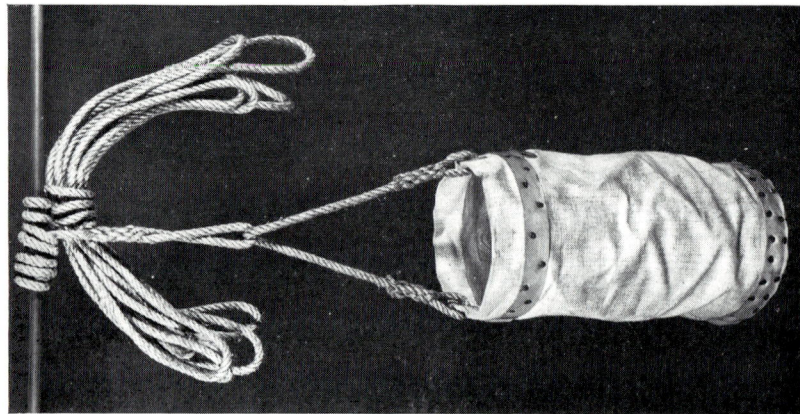

FIGURE 12.

The rubber sea-temperature bucket Mk 3B

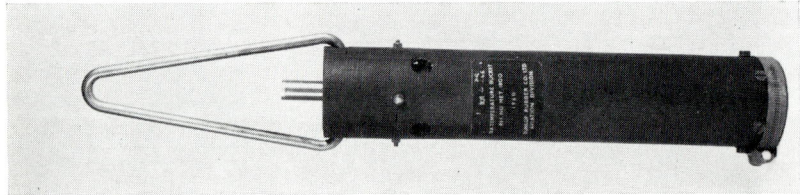

(b)

FIGURE 11.

The rubber sea-temperature bucket MET 1800

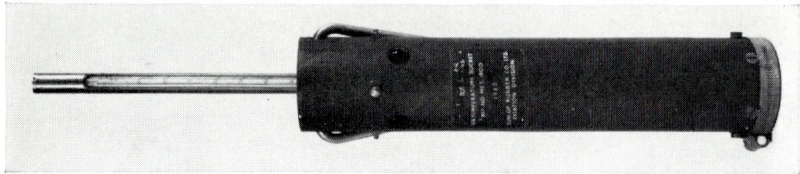

(a)

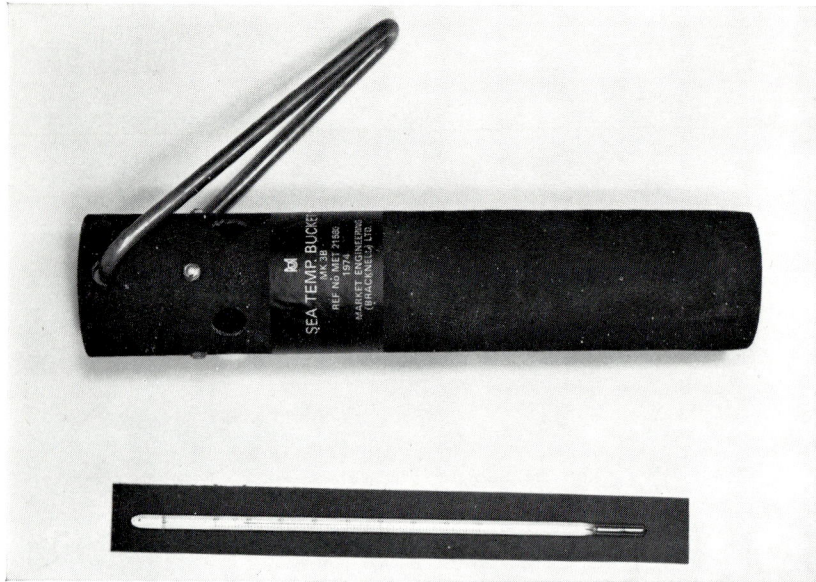

FIGURE 10.

The canvas sea-temperature bucket

(b) View with covers removed

(a) Assembled for flight

FIGURE 13. The British radiosonde Mk IIb

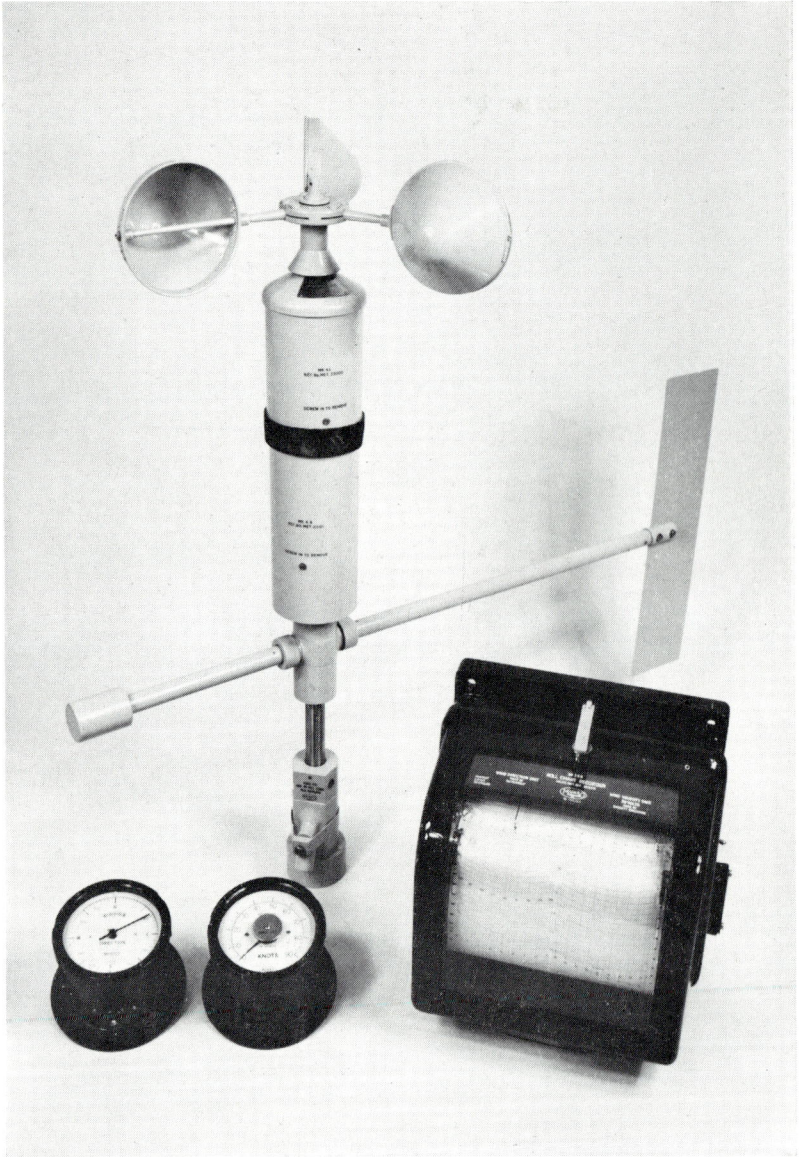

FIGURE 14. Cup generator anemometer Mk 4 with wind vane, dials and recorder

FIGURE 15. Cup generator anemometers and wind vanes in an ocean weather ship

FIGURE 16. Solarimeter (pyranometer) on board an ocean weather ship. The front of the mounting is open and shows the silica gel container connected by a tube to the solarimeter. The pendulum part of the gimbal mounting is also seen below. The white horizontal guard plate (305 mm in diameter), which is normally fitted around the solarimeter in the plane of the base of the glass dome, has been removed for the purpose of this photograph.

The socket having been screwed to the bulkhead, the instrument should be carefully lifted, the hinged part of the suspension arm bent back at right angles and shipped into the socket so that the longer portion of the suspension arm is horizontal. The mercury should then fall gradually and the instrument will be ready for observation in about two hours; this also allows time for the instrument to take up the temperature of its surroundings. Sometimes in a new tube the mercury does not readily quit the top of the tube. If, after an hour or so, the mercury has not descended, tap the cistern end rather sharply, or make the instrument swing a little in its gimbals, which should cause the mercury to fall in the tube. If this method does not succeed, the force of the tap must be slightly increased, but undue violence must not be used.

Figure 4 shows a marine barometer (a) housed in its case when in harbour and (b) in sea position. The case should be firmly secured to the bulkhead. The socket is screwed near the bottom of the case and a clip 'A' is provided to hold the barometer in its housed position when in port. (Note.—the short screws holding the socket in the box are insufficient to hold the weight of the barometer when in the sea position. These short screws should be replaced by longer ones at least 4 cm, screwed through the socket and box and into the bulkhead. A hook should be fitted to secure the lid open while at sea.

Taking down the barometer. Whenever a barometer has to be unshipped and placed in its box, first lift the instrument out of its socket and bring it gradually into an inclined position to allow the mercury to flow very gently up to the top of the glass tube, avoiding any sudden movement which would cause the mercury to strike the top of the tube with violence. The absence of air in the tube makes the force of the blow little different from that of a solid rod of metal, so that it might break the tube. The barometer should then be taken lengthwise and laid in its box. To be carried with safety it should be held with the cistern end upwards or lying flat and it must on no account be subject to jars or concussions, which might cause air to find its way into the upper end of the tube, even if they did not damage the instrument.

Care of the barometer. The barometer should be kept clean and dry. The gimbal screws should be examined occasionally, as they are usually made of brass and in the course of time may wear through, owing to the movement of the instrument at sea, particularly in small ships. Dust, particularly on the correction slide, should be removed by gently brushing the instrument with a camel-hair brush or a soft cloth. Metal polish should *never* be applied. A very little clock oil or log oil may occasionally be used for lubrication.

If the rack and pinion of the correction slide become very stiff they may be overhauled as follows. Remove the slide from the barometer and place it face downward. A small brass block securing the pinion in position will then be seen. Remove this by taking out the four screws. Wipe the pinion and its bearing with a soft rag to remove dirt and old oil; apply a little fresh clock oil. Now remove the four small screws, two at each end of the rack. The slider can then be taken out. Wipe off all dirt and old oil from the rack and bearing surfaces. Put a drop of fresh clock oil on the rack and on the back of the slider. Reassemble, taking care to see that the pinion is properly engaged in the rack before tightening the screws. The slider should then move up and down quite smoothly.

The screws which secure the latitude scales should not be touched during this operation.

Temperature of the instrument. This is read to the nearest whole degree on the scale of the attached thermometer. The observation should be made immediately on reaching the instrument in order that the thermometer should be affected as little as possible by bodily heat radiated from the observer.

Height of the mercury column. After the temperature of the barometer has been read, the barometer may be touched with the hand, but care should be taken to do this as lightly as possible. Tap gently with the finger until the tapping no longer affects the shape of the mercury surface in the tube. Turn the milled head at the side of the instrument until the lower edge of the vernier and the lower edge of the sliding piece at the back of the instrument, which moves with the vernier, when in line, appear just to touch the uppermost part of the domed surface of the mercury.

A white background, e.g. a piece of white paper placed behind the instrument, is an advantage. An electric torch can be used to illuminate the background but do not use a naked light as this may lead to an inaccurate setting.

If the mercury is not perfectly pure, it may happen that, when the barometer is falling, the top of the mercury column no longer shows a domed (convex) surface. The surface may be flat or even concave. The exact setting of the vernier is much more difficult under these conditions. The observer should move the vernier slowly downwards, keeping its lower edge and the lower edge of the sliding piece at the back in line as well as he can, till the white background just disappears at the centre of the tube. He should then move his eye a little up and down to make sure that the white background still remains invisible at

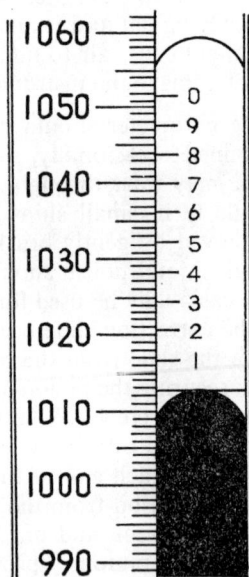

Reading 1012·7mb

FIGURE 17. Reading a millibar barometer, using a vernier scale. The reading is 1012·7 mb

the centre of the tube, before he takes the reading. In this case a bright white background, for example a piece of white paper held behind the top of the mercury column, is almost essential.

Figure 17 illustrates the process of reading the vernier, which is read in the same way as that of a sextant. The illustration shows a millibar barometer, the reading being 1012·7 millibars.

Accuracy of reading. It is important that barometer readings for transmission by radio or for entry in a meteorological logbook be carefully made. An erroneous pressure reading may considerably mislead a forecaster, and the error may have serious consequences for ships at sea, particularly in tropical storm areas. Some possible sources of error are shown below.

ERROR DUE TO THE CORRECTION SLIDE BEING IMPROPERLY SET. In all cases the correction slide (see page 7) should be reset before reading the barometer. This is particularly important where the ship is changing her latitude.

ERRORS DUE TO PARALLAX. If the eye is not in line with both the bottom of the vernier and the sliding piece at the back, the reading will be incorrect owing to errors of parallax. Whether the eye is too high or too low, the reading will be too high. If the eye is too high, only the front of the vernier can be seen and this will be in line with the top of the mercury column and the eye. If the eye is too low, the sliding piece at the back of the vernier will be in line with the top of the mercury column and the eye, the lower front edge of the vernier being indistinguishable. (See Figure 18.)

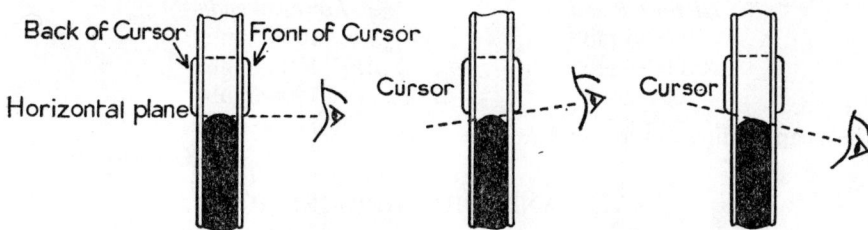

FIGURE 18 Errors due to parallax

Correct position of the observer
The eye of the observer and the lower edges of the front and back of the cursor are all in a horizontal plane tangent to the meniscus.

Incorrect position of the observer
In neither case is the eye of the observer in the horizontal plane tangent to the meniscus, and the result in both cases is too high a reading.

ERRORS IN READING THE MAIN SCALE. The simplest error that can be made in reading the barometer is that of making an actual mistake of 10 mb or 1 mb; such an error is usually due to making a mistake in counting the number of divisions on the fixed scale. The only means of guarding against such errors is care. After a reading has been logged it should be checked to make sure that no misreading has been made. In making the first reading, attention should be concentrated on the accuracy of the last figure (tenths of a millibar); in the check reading attention should be concentrated on the figures of higher value.

ERROR DUE TO WIND. It has been found that strong winds, blowing near the barometer, may affect the indicated pressure. When such winds are blowing during a observation, therefore, the doors of the charthouse in which the barometer is hung should be kept closed. This applies just as much to the lee door as to the weather door.

11

ERROR DUE TO CHANGE OF DRAUGHT. A change in height above the water-line after loading or unloading, is sometimes overlooked. In some ships the difference in pressure readings between the light and loaded draught could approach 1·5 mb.

ERRORS DUE TO PUMPING. When a ship is in a seaway, the mercury of the barometer may oscillate up and down in the barometer tube. This is termed 'pumping' and is due to the following causes:

(a) Oscillations of the mercury caused by the pitching and rolling of the ship.
(b) Oscillations caused by the swinging of the instrument about its point of support.
(c) The effect of wind gusts on the air pressure of the room in which the barometer is hung.
(d) Variations of atmospheric pressure caused by the change of height of the ship due to her vertical motion on the waves.

The mean reading should be recorded, to obtain which the vernier should be set by eye midway between the highest and lowest positions of the mercury column. The pumping is often very irregular, and in order to get an accurate mean the observer is advised to take three pairs of readings, one of each pair being the highest reading obtainable and the other the lowest. The result recorded is the mean of the whole set. Thus, if observations obtained were as follows:

Highest Reading				Lowest Reading
1007·6 mb	...	...	...	1006·5 mb
1007·5 mb	...	...	...	1006·6 mb
1007·7 mb	...	...	...	1006·6 mb

the mean reading would be 1007·1 mb.

THE ANEROID BAROMETER

The aneroid barometer consists of a circular metallic chamber exhausted of air and hermetically sealed. Variations of atmospheric pressure produce variations in the dimensions of the vacuum chamber and these changes are magnified mechanically, optically or electrically so that the atmospheric pressure may be read on a convenient scale. The principle of the aneroid barometer was first suggested in 1698 but no useful instrument was constructed until 1843.

The majority of aneroid barometers indicate the pressure by means of a pointer which rotates around a graduated dial. The vacuum chamber, usually called the aneroid capsule, has to provide the force needed to move the pointer and this prevents it from responding freely to pressure variations. This type of instrument is useful in showing pressure changes and some of the better quality instruments are suitable for all pressure readings. The aneroid has the advantage that, unlike the mercurial barometer, it does not suffer from 'pumping' although it does rise and fall slightly with change of height of the ship in the waves of a seaway.

Precision aneroid barometer. The Meteorological Office has now adopted, as its standard instrument, a type of precision aneroid barometer in which the force required to operate the indicating mechanism is provided by the observer, allowing the capsule to respond freely to pressure changes. The sensing element is a stack of three disc-type aneroid capsules fixed to the inside wall of a cast metal box. Some magnification of the capsule movement is provided by a lever, pivoted on jewelled bearings. One end of the lever is kept in contact with the capsule by means of a light hairspring and a micrometer screw, which extends through the case and actuates a digital counter, is brought into contact with the other end of the lever by the observer. Contact is indicated by a small cathode-ray tube; a continuous line of light indicates that contact is made and a broken line of light indicates that the circuit is broken. When the micrometer screw is set so that the contact is just broken the digital counter indicates the pressure in millibars and tenths. The box containing the aneroid capsules is completely air-tight except for one hole, and that orifice is fitted with a damping device which restricts the response of the instrument to the rapid pressure variations caused by the rise and fall of a ship.

The Precision Aneroid Barometer Mk 2 (as shown in Figures 6a and 6b) is the type issued to Voluntary Observing Ships.

INSTALLATION OF THE MK 2 ANEROID BAROMETER. The installation of the barometer on board ship should be carried out by the Port Meteorological Officer. It should be mounted on a mounting plate on a fore-and-aft bulkhead. When this has been done the damping cap should be fitted. Firstly unscrew the Static Vent (Figure 6a) reverse it and screw it back in finger-tight making sure that the O ring on the static tube beds firmly.

READING THE MK 2 ANEROID BAROMETER. Press the black switch button. If the thread of light in the Cathode Ray Indicator is broken, turn the knob so that the pressure reading decreases until the thread becomes continuous. When the light is continuous turn the knob so that the pressure reading increases until the thread of light just breaks. This should be repeated a second time to avoid any error due to overshooting. The pressure should be read in the window when the light breaks. If parts of two figures show equally in the tenth-of-a-millibar position the odd number should be taken.

The pressure as read must be corrected to mean sea level. Firstly apply the correction from the calibration correction card supplied with the barometer and secondly apply the correction given by Table 10. This must be done for all observations.

MAINTENANCE OF THE MK 2 ANEROID BAROMETER. The only maintenance required is the renewal of the battery. When the indicator thread becomes dim and it is difficult to see whether or not it is broken the battery should be changed.

REMOVAL OF BATTERY. The battery cover is a plastic disc on the right-hand side of the barometer. Press it gently inwards and rotate it slightly anti-clockwise and it can be removed. Lever the spring clip from the top contact of the battery and remove the battery, insert a new one and then push the spring clip on to the terminal. Replace the plastic cover by pressing in gently and rotate it slightly clockwise. The battery should last for about 12 months.

Corrections to aneroid readings. Aneroid barometers of good quality are compensated, by the manufacturers, for such changes in temperature as they are likely to experience, either by leaving a calculated small amount of air in the vacuum chamber, or by the use of a bimetallic lever. Such aneroids, therefore, do not require correcting for temperature. No aneroids require correcting for latitude, as the principle on which they are based is the balancing of atmospheric pressure by the elasticity of metal, so that the force of gravity does not come into the picture. The only corrections which should be applied to an aneroid reading are those for altitude (see Tables 10 and 11) and where necessary for index error.

13

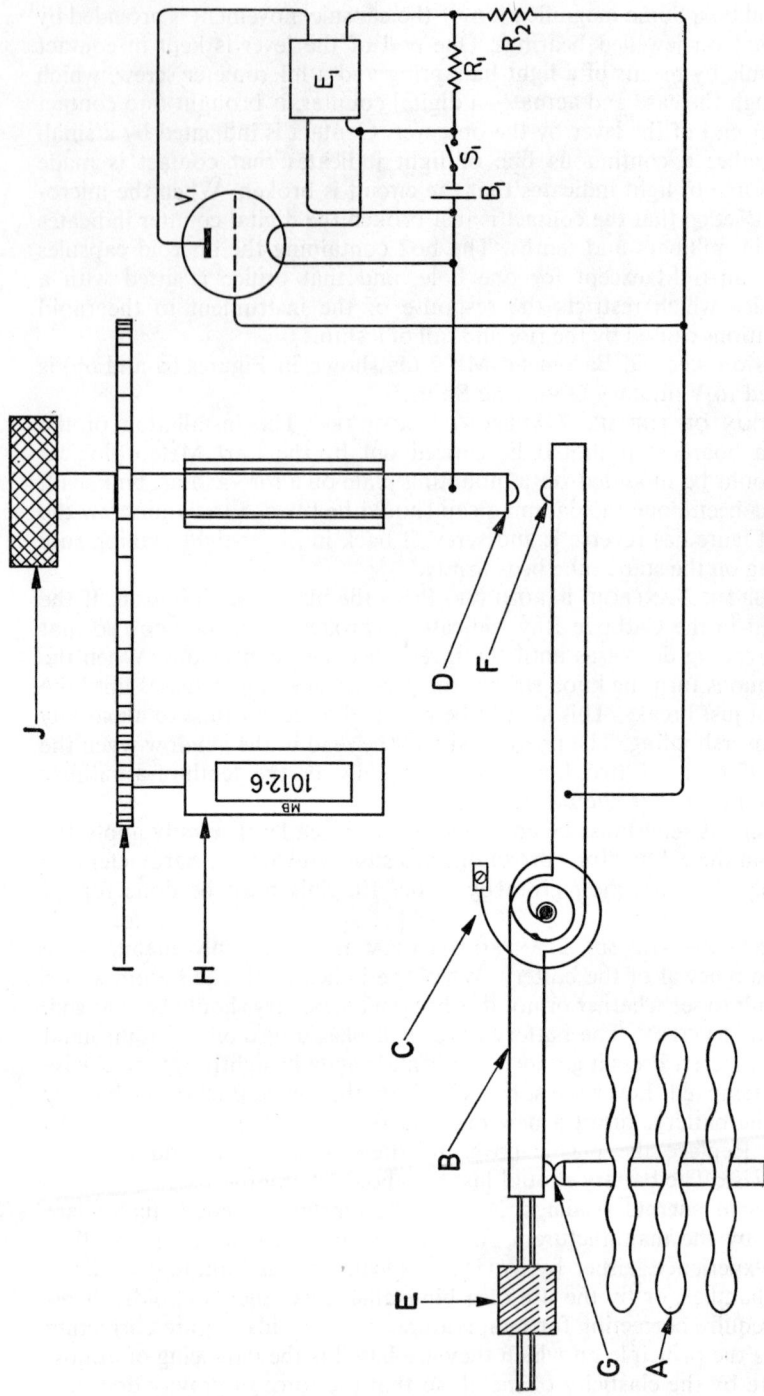

FIGURE 19. Schematic drawing of Precision Aneroid Barometer Mk 2

A	Aneroid capsule assembly	E	Counterbalance	I	Gearing
B	Pivoted bar	F	Sliding electrical contacts	J	Knob
C	Hairspring	G	Mechanical contacts	V_1	'Magic-eye' indicator type DM70
D	Micrometer-type spindle and nut	H	Counter	S_1	Switch

B_1	Battery, 1·5 V
E_1	Voltage converter
R_1	Resistance
R_2	Resistance

14

Precautions necessary with an aneroid barometer. The instrument should be placed where it is not liable to sudden jars which may alter its index correction, rapid changes of temperature and where the sun cannot shine directly on to it. Dial aneroids should be tapped gently before a reading is taken as the pointer is liable to stick. This is not necessary with the digital precision aneroids.

All aneroids require careful comparisons with barometers whose accuracy can be relied on, as changes in the elasticity of the metal of which the vacuum chamber is composed may cause appreciable variations in the index correction. Such changes are rare in good quality instruments. Every opportunity should be taken when the vessel is in harbour of making a comparison with a reliable barometer. With a Precision Aneroid Barometer Mk 2 the damping cap should be removed before making a comparison with a check barometer. Comparisons against a mercury barometer while the ship is at sea are not likely to be satisfactory as the readings of a mercury barometer on a rolling ship cannot be considered as reliable for this purpose.

All Port Meteorological Officers and many harbour and mercantile marine offices have a standard barometer which is available for such comparisons. A record should be kept of all barometer comparisons; this will be useful in assessing the reliability of the instrument and the correction to be applied to dial aneroids when at sea.

Adjustment of aneroid readings. The reading of a dial aneroid may be corrected, if desired, by means of the adjusting screw at the back. Whenever such an alteration of the index correction is made, the fact should be noted, with the date. Such adjustments should, however, only be made if the index correction becomes too great; small changes in the index error of the instrument should be allowed for by altering the correction to the applied readings. No attempt should be made to alter the setting of the Meteorological Office digital Precision Aneroid Barometer Mk 2.

THE BAROGRAPH

A typical barograph is shown in Figure 5. It is constructed on exactly the same principle as the aneroid barometer, but records its readings by the movement of a pen over a suitable chart.

To increase the movement through which the pen travels in response to pressure changes, the vacuum chamber takes the form of either a number of individual capsules or one large chamber with corrugated walls. The bottom of this vacuum chamber is anchored to the instrument base while the top is connected to the pen arm through a series of levers which still further magnify the movement by pressure changes and which can be arranged to compensate for temperature changes. The variation of volume of the vacuum chamber is thus translated into a vertical movement of the pen arm. This pen arm carries a pen filled with special ink and pressure changes are presented as a mark on a chart attached to a clockwork-operated drum. This pressure is recorded as a continuous line whose height at any point represents the pressure at the time it was recorded. This record is known as a barogram.

The barograph is a valuable adjunct to the barometer aboard ship in providing a continuous record of atmospheric pressure between the times at which the barometer was read.

15

The barograph is not a precision instrument and should never be used as an alternative to the barometer for measuring atmospheric pressure at fixed times. Its advantage is that it provides a graphical record of fluctuations of pressure, together with the times at which they occur, such as the moment of passing of a line squall and its readings are valuable to the meteorologist and to the mariner for various practical purposes.

The open-scale barograph. Barographs may be made in various scales of size, the smaller being more convenient when space is limited. In such smaller instruments the vacuum chamber and the clock drum are themselves smaller, and the pressure changes recorded on the chart are of a correspondingly reduced scale. To report barometric tendencies with the accuracy required for synoptic meteorological observations, it is desirable that larger, and hence more open scaled barographs be used. Records from such barographs, when carried on board ships, may, however, be unsatisfactory because, due to their greater sensitivity, the trace is not a fine line but a ribbon of appreciable width, resulting from vibration, pressure changes from gusts of wind and from the movement of the ship. Because of this, an oil-damped open-scale barograph (Meteorological Office Marine Mk 2) is used aboard British Selected Ships. In this instrument the vacuum chamber is contained in a brass cylinder filled with oil which, to compensate for changes of volume of the vacuum chamber, must pass through a small orifice. In this way short time-period changes are damped out and only the major persistent changes shown on the chart. As a further precaution against vibration the instrument is mounted on rubber pads.

Care of the barograph. The barograph is a delicate instrument and must be handled carefully. Friction between the working parts of the apparatus must be avoided as far as possible. The bearings should be cleaned occasionally and oiled with good clock oil, care being taken to remove excess of oil.

Friction occurs between the pen nib and the paper on which it writes. The pressure of the pen nib on the paper should be reduced to the minimum consistent with a continuous trace; this pressure should be tested from time to time.

In the open-scale barograph now in use the pen arm which carries the pen nib is suspended like a gate and it is so arranged that the slope of the gate bearings is adjustable. It is thus possible to regulate the pressure of the pen nib on the chart. In Figure 5, A denotes the gate suspension which is suitably adjusted before issue.

Excess of ink in the pen nib should be avoided. Do not let ink come in contact with the metal pen arm which carries the pen nib as this will cause the pen to adhere firmly to the pen arm so that it cannot be removed and cleaned. The ink may also cause the metal to become brittle and break. Should the pen arm become inked, it should be washed and slightly oiled. A thin, clear trace on the chart should be aimed at. The pen nib should be washed from time to time in water or methylated spirit. The point of the pen nib should be fine, so as to give a narrow trace, but it must not be so fine as to scratch or stick to the paper. A new pen nib may be improved by drawing the point once or twice along an oil stone, but any oil should afterwards be removed. The pen nib should be filled with ink weekly, using the ink-bottle and filler supplied.

The barograph, when used on board ship, should be located in a position where it will be least affected by concussion, vibration or movement of the ship.

Setting of the barograph. The barograph is set to give the correct mean-sea-level reading by comparison with the reading of the mercury barometer or precision aneroid, after the latter has been corrected to give mean-sea-level pressure.

In the type of barograph shown in Figure 5, the setting is made by adjusting the height of the fulcrum of the principal lever B by means of the milled head screw C on the central bridge. In other instruments the adjustment is made by raising or lowering the point in the base plate to which the lowest of the set of aneroid boxes is fixed. This is done by means of a milled-head screw on the base plate near the aneroid boxes.

Standardizing the barograph. Like the aneroid barometer, and for the same reason (the possibility of changes in the elasticity of the metal of which the vacuum boxes are composed), the readings of the barograph should be compared at least once a week with those of a mercury barometer, duly corrected. The most suitable time is when the weekly chart is changed, and the reading of the barometer, together with the date and Greenwich Mean Time (GMT), should be entered up on the chart. If the ship does not carry a barometer, every other opportunity of making such a comparison should be taken.

Adjustments to the barograph should not be made too frequently, but only if its readings become appreciably different from those of the barometer, and a note of the adjustment should be made on the chart, giving time and date.

The barograph clock and chart. The barograph may be fitted with various clocks which will rotate a drum quickly or slowly as desired, round which the chart is fixed. For many applications a rotation in 24 hours will be necessary to show up the small-scale features of pressure changes but, for most uses at sea, the clock chosen will rotate once per week. The chart must therefore be changed weekly, the clock being wound at the same time. Before removing the chart from a small barograph the pen arm must first be moved away from the chart by means of the lever provided. On open-scale barographs the pen arm automatically lifts off the chart when the lid is lifted. Before the new chart is put on the drum, the date and time should be entered on it in pencil. Time marks should be made each day at 1200 GMT and just before the chart is removed, the times being entered on the chart, for the purpose of correcting the time scale should the barograph clock run fast or slow. The barograph should be kept to GMT throughout the voyage. For the purpose of making time marks, barographs have a small button on the outer case which, when depressed, acts on a rubber roller (D in Figure 5), which slightly moves the pen arm vertically.

Before fixing the chart on the drum, the latter must first be lifted from the clock by removing the key and unscrewing the milled nut which holds the drum in place. The chart is then placed round the drum where it is held in position by two short spring clips that hold its bottom and top edges. When fixing on the drum, care must be taken that the horizontal lines printed on the chart are parallel to the flange at the base of the drum. As the length of the chart is slightly greater than the circumference of the drum, there is some overlap when the chart is put on the drum. The last portion of the chart should come on top of the first portion, so that if the chart is not changed at the end of seven days, the pen will not catch on the edge of the chart and tear it, or damage itself.

The drum is then replaced on the clock and the whole is rotated till the pen records the correct GMT. In order to avoid time errors that might be caused by backlash in the teeth of the clock gears, the final movement of the drum, when setting it, should be in the opposite direction to that in which it normally rotates.

17

THE CHANGE OR TENDENCY OF THE BAROMETER

The change or tendency of the barometer, always a valuable observation to seamen, is also of considerable value to the forecaster.

The barometric tendency, by international usage, is defined as the change in the barometric pressure in the last three hours. It is required in radio weather messages and is read off from the barograph. The position of the pen on the chart at the time of observation, and the reading of the trace three hours earlier, should be noted, if possible to the tenth of a millibar. The difference between these two readings will give the tendency. It should not be taken as the difference between two readings on the barometer, but should always be read off from the barograph, since the barograph method is less liable to error, and anyway the barometer is not customarily read every three hours at sea. Also, mistakes in reading a barograph are more likely to be detected, owing to the continuous availability of the trace.

It is essential that the barograph trace should be fine and sensitive, with the instrument free from mechanical faults such as sticking, and as far as possible not vitiated by the effects of vibration, or of unequal heating due to sunshine or nearby sources of heat.

Allowance for course and speed. To estimate the true tendency of the barometer reported from a ship under way, a meteorological service needs to allow for course and speed, and, therefore, in a ship's weather report provision is made for reporting the course and speed of the ship. This allowance for the course and speed of the ship should *not* be made by the observer on board ship when reporting tendency in a weather message. This allowance can be readily made at the meteorological office ashore when the observations are studied by forecasters or processed by computers.

The characteristic of the barometric tendency. This is the name given to the coded description of the nature of the changes the pressure has undergone in the last three hours. It is generally required in ships' weather reports, and is read off from the barograph trace. The diagrams in Figure 20 show the various pressure changes that might have to be reported, together with the code figures to be used in reporting them.

The codes to be used in reporting the barometric tendency and characteristic are given, with other codes, in the *Ships' Code and Decode Book* (Met. O. 509) and in the *Admiralty List of Radio Signals*, Vol. 3.

The diurnal variation in the pressure. Superposed upon its irregular variations due to changes in the weather, the barometric pressure has a regular rise and fall twice a day, the maximum values occurring at about 10 and 22 hours and the minimum values at about 04 and 16 hours, local time. In temperate regions the amplitude of these diurnal variations is comparatively small, so that they are usually lost in the much greater irregular variations of these regions, but nearer the tropics, the amplitude of the diurnal variation increases and the magnitude of other changes in general decreases, so that the diurnal variations become very marked and can be clearly seen, day after day, on a barograph chart. In these regions, therefore, barometric changes do not indicate changes in the weather, unless they remain considerable *after the diurnal variation has been discounted.*

Tables have been prepared for the Atlantic, Pacific and Indian Oceans, between latitudes 0° and 20°, N or S, showing the corrections for diurnal variation to be applied to the observed pressure to reduce it to the mean for the day, and the average values of the barometric change in an hour, throughout

Code Figure	Trace	Description of Curve	Pressure *now*, compared with 3 hours ago
0		Rising, then falling Rising, then falling	The same Higher
1		Rising, then steady Rising, then rising more slowly	Higher
2		Rising, (steadily or unsteadily)	Higher
3		Falling, then rising Steady, then rising Rising, then rising more quickly	Higher
4	—	Steady	The same
5		Falling, then rising Falling, then rising	The same Lower
6		Falling, then steady Falling, then falling more slowly	Lower
7		Falling (steadily or unsteadily)	Lower
8		Steady, then falling Rising, then falling Falling, then falling more quickly	Lower

FIGURE 20. Code numbers used to indicate the characteristic of barometric tendency

the day, due to the diurnal variation. These tables are given in the meteorological text of the appropriate Admiralty sailing directions. Corresponding figures do not differ greatly from one ocean to another or between north and south latitudes and have been averaged in this handbook to give values that will be approximately correct in any ocean for the two bands of latitude 0°–10°, N or S, and 10°–20°, N or S. These values are shown in Tables 12 and 13 (see also Figure 21). In the tropics, should the barometer, after correction for diurnal

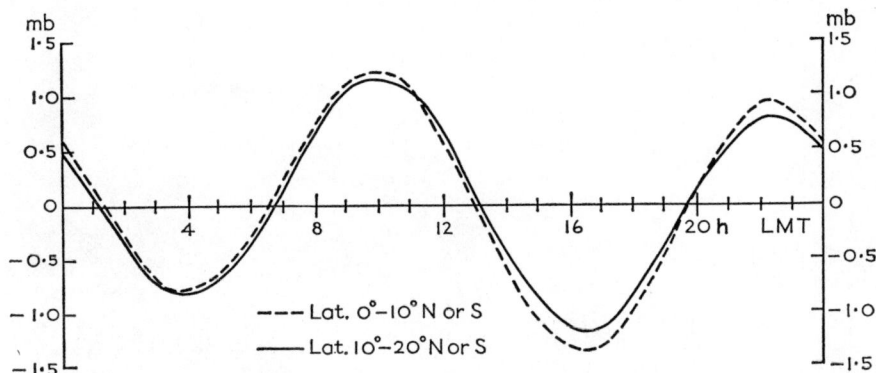

FIGURE 21. Mean diurnal variation of pressure

19

variation (Table 12) be as much as 3 millibars below the monthly normal for the locality, the mariner should be on the alert, as there is a distinct possibility that a tropical storm has formed, or is forming. A comparison of subsequent hourly changes in his barometer with the corresponding figures in Table 13, will show whether these changes indicate a real further fall in pressure and, if so, its amount.

When the observer on board ship is reporting barometric tendency, or entering it up in his log, he should not *correct it for changes due to normal diurnal variation.* This correction, like the correction for course and speed of the ship, is made, if necessary, as a matter of routine by the meteorological office receiving the observations.

CHAPTER 2

The Measurement of Temperature and Humidity

The temperatures normally measured at sea for meteorology are those of the air, at the height of the bridge, and of the sea, just below its surface. Humidity, i.e. a measure of the evaporated water contained in the air, is also required, but as this is obtained by similar instrumentation, it is included under the same general heading.

Thermometers. Any device which measures temperature is a thermometer. There is a wide range of physical phenomena related to temperature, almost any one of which may be used as a thermometer but the two which will be most frequently encountered are based on the expansion of a suitable substance with increased temperature and, similarly, the change of electrical resistance in a conductor. The simplest, cheapest and most commonly encountered device is the liquid-in-glass thermometer, the liquid employed being mercury or alcohol. Such a thermometer consists of a glass tube of very fine bore, at the end of which a bulb has been blown to act as a reservoir. The whole of the tube and bulb is filled with the chosen liquid at a high temperature and the open end of the tube is then sealed. On cooling, the liquid will contract until the tube is only partly filled by liquid, the exact point reached by the liquid being a measure of the temperature of the thermometer, and hence, under suitable conditions, of its surroundings, at any given moment. A scale may now be engraved on the tube, or thermometer stem, to allow actual temperature to be read.

The thermometer was invented at approximately the same time as the barometer. Galileo made a crude kind of thermometer in which the liquid was open to the air. True thermometers were first brought into general use by the Grand Duke Ferdinand II of Tuscany who is said to have possessed such instruments in 1654. The liquid used in these early thermometers was alcohol.

While mercury is the most satisfactory liquid for general thermometric use, thermometers intended for very cold climates contain pure alcohol. The reason for this is that mercury would solidify at the low temperatures of polar regions. Mercury freezes at about $-39°C$ ($-38°F$) while alcohol freezes only at $-130°C$ ($-202°F$), though it becomes a thick liquid and therefore useless for thermometric purposes at $-90°C$ ($-130°F$).

Thermometers employing the electrical change of resistance due to temperature give no direct visual indication but must be placed in an electrical circuit which will enable the resistance and hence, from previous calibration, the temperature, to be measured. Such thermometers usually are constructed from a length of fine wire, drawn from a material such as platinum, tungsten etc., which is ductile and will not corrode with time. For meteorological use, a spool of such wire is permanently enclosed in a small-diameter metal cylinder, for protection.

Graduation of thermometers. The earliest known graduation of a thermometer was that made in 1701 by Sir Isaac Newton, who divided the range of temperature between the freezing point of water and the temperature of the human body into twelve degrees.

21

Later scientists used as fixed points the temperature of a mixture of salt and ice, and the boiling point (at standard pressure) of water. The SI unit of temperature is the kelvin (symbol K), widely used in scientific work. It is defined as the fraction $1/273 \cdot 16$ of the thermodynamic temperature of the triple point of water; the triple point of a substance being the pressure–temperature condition, unique for a given substance, at which the substance may exist in the solid, liquid or gaseous state. On this scale water freezes at $273 \cdot 15$ K and boils (at standard pressure) at $373 \cdot 15$ K. For normal use the Celsius (known as the centigrade) temperature (symbol °C) has also been approved by the International Committee on Weights and Measures. It is defined as:

$$t = T - T_0$$

where t = Celsius temperature, T = thermodynamic temperature in kelvins, and $T_0 = 273 \cdot 15$ K. On this scale water freezes at 0°C and boils (at standard pressure) at 100°C. It is the official scale for the measurement of all meteorological temperatures.

The Fahrenheit scale (symbol °F), once in general use in the English-speaking world, now rapidly becoming obsolete, had two fixed points; zero was taken as the temperature of a mixture of salt and ice and 100 as the temperature of the human body. This gave the freezing point of water as 32°F and the boiling point (at standard pressure) as 212°F. For a time the Meteorological Office used a scale of temperature called 'Absolute' which approximated to the scale now called the kelvin. The symbol was °A. On this scale the water froze at 273°A and boiled (at standard pressure) at 373°A. This scale may still be found on older mercury barometers and in some correction tables.

Conversion of thermometer scales. To convert Celsius readings to Fahrenheit use the following rule: Multiply by 9/5 and add 32. Similarly, to convert from Fahrenheit to Celsius, subtract 32 and multiply by 5/9. From Fahrenheit to kelvin (formerly known as Absolute), proceed as for Celsius and add $273 \cdot 15$. Table 20 gives the values on the Celsius and kelvin scales corresponding to each degree Fahrenheit, from 0°F to 119°F.

Scale markings. Thermometer scales can take various forms. As the liquid-in-glass thermometer is particularly fragile, it is usually protected in some way or other, and the scale is often incorporated in this protection. In the standard Meteorological Office sheathed thermometer (Figure 8a) the scale is engraved directly upon the thermometer stem, the back being whitened to allow easy reading. The thermometer stem is then enclosed in an outer glass tube, which adds to the strength of the whole and protects the scale from erosion. In this case the thermometer is therefore read through the outer glass tube. In other instances the basic glass thermometer is partially enclosed with a wood and metal protecting frame, to which is attached a separate scale, engraved on a flat white surface. In such cases the thermometer itself will carry a more simple scale, without numbers to indicate the true temperature. This is done to ensure that the thermometer has not moved with respect to the fully engraved scale, and this point should be frequently inspected when such thermometers, now rather obsolescent, are used.

Electrical thermometers are usually read by dial and pointer, the pointer either being operated by the resistance thermometer electrical circuits or by manually setting to achieve a prescribed effect—the 'zeroing' of a secondary pointer, or by the lighting or extinction of electric lamps.

Reading the thermometer. The thermometer should be read with care. Though the Fahrenheit scale is graduated only to whole degrees,* the reading on both Celsius and Fahrenheit thermometers should be given by estimation to the nearest tenth of a degree. This is not only necessary for general accuracy but also for practical reasons, i.e. the computation of relative humidity and the dew-point (see page 26), and the determination of the difference between air and sea surface temperatures. In some coded radio weather messages, however, the temperature is required only to the nearest degree.

When reading a thermometer, care should be taken to keep the eye at the same level as the end of the column, otherwise there will be an error due to parallax.

The mercury column of a thermometer occasionally separates in one or more places. The thermometers should therefore be examined before each observation to see if the column is continuous. If there is any break in the column, take the instrument down, swing it briskly at arm's length with the bulb end away from you till the column is again continuous, and replace it. After this, give the thermometers another 10 minutes to pick up the correct temperatures again, before taking the observation. With the alcohol-in-glass thermometer some alcohol may flow into the upper end of the tube unless the thermometer is stored with the bulb end downwards.

Thermometers should be kept clean. In damp weather any moisture should be removed from the dry bulb a little while before taking the reading. The graduations on the glass of mounted thermometers may in time become indistinct. Since the marks are cut in the glass, a rub with an ordinary lead pencil or a wipe over with Indian ink will make the graduations clear again.

For all types of liquid-in-glass thermometer it is best always to store in a vertical or near vertical position and never with the bulb higher than the end of the stem. In consequence a spare thermometer is most conveniently retained in its box, which can be conveniently located in a rack or a clip which will hold it in a vertical position.

THE DRY- AND WET-BULB THERMOMETERS

An instrument for measuring the humidity of the air is called a hygrometer. There are several kinds of hygrometers, but the form in common use, the dry- and wet-bulb thermometers, also known as Mason's hygrometer or a psychrometer, is the simplest and is described below.

Of the two thermometers contained in the thermometer screen, no more need be said of the dry-bulb for measuring air temperature, beyond ensuring that it is secured firmly into the clips provided. The operation of and attention to the wet-bulb thermometer requires a little further description.

Operation of a wet-bulb thermometer. The evaporation of water requires heat, the 'latent heat of evaporation'. This is derived from the surroundings—the air, the water itself and/or from the thermometer used for the measurement. The faster the evaporation the greater the demand for latent heat and hence the greater will be the cooling of the surroundings. However, the rate of evaporation under any particular circumstance will be determined by the dryness of the surrounding air, the air temperature and the rate at which air flows past the

---

*Ships' Celsius thermometers are graduated in half degrees.

thermometer. A measure of humidity, i.e. the degree of dryness or wetness of the air, can thus be obtained by wetting a thermometer and noting the degree to which it is cooled. In practice it is both inconvenient and unnecessary to wet the whole thermometer—wetting the bulb alone will suffice. The bulb itself is thus enclosed in a small muslin bag, tied on by means of a wick, the other end of which is placed in a small water container placed beside the thermometer. Capillary action will then ensure that the muslin is kept wet, and the cooling action of evaporation can then be measured by reading, firstly, the dry-bulb thermometer, then the wet-bulb thermometer and by subtraction, the difference, the 'wet-bulb depression'. The third requirement is knowledge of the rate of the air flow. A value has been assumed so that, from the air temperature and wet-bulb depression, a reasonably correct measure of humidity may be obtained from the tables provided. This combination of a dry- and wet-bulb thermometer is known as a psychrometer.

Air can contain only a limited amount of evaporated water, according to its temperature. When this point is reached, no further evaporation will take place and the wet-bulb thermometer will read the same as the dry-bulb thermometer. The air is then said to be saturated. If the air becomes drier, the rate of evaporation increases and the wet-bulb temperature falls. The depression of the wet-bulb can reach over 20°C (36°F) in a hot dry climate, such as that of Khartoum during part of the year. It sometimes amounts to 10°C (18°F) in England, but at sea the difference seldom reaches 5°C (9°F). When the humidity of the atmosphere is high, during or just before or after rain, when fog is prevalent, or when dew is forming, there is little or no evaporation and the two thermometers give the same, or very nearly the same reading.

We may sum up the facts about humidity and the dry- and wet-bulb thermometers as under:

Humidity		Evaporation		
High	...	Weak	...	Dry and wet bulbs read almost the same.
Low	...	Intense	...	Wet bulb reads much lower than dry.

Muslin and wick for wet bulbs. The wet-bulb thermometer needs careful attention in order to get correct readings. The bulb of this thermometer should be covered with a single thickness of thin *clean* muslin or cambric, which is kept moist by attaching to it a few threads of darning cotton dipping into the small reservoir of water placed near it.

From the muslin provided, a small piece should be cut, sufficient to cover the bulb, and should be stretched smoothly over it, creases being avoided as far as possible. The muslin is kept in place by attaching the cotton wick in the following way. Take a round turn in the wick, with the strands middled on the bight, and pass the ends through the bight, forming a round turn and cow hitch. Any superfluous muslin or loose ends should then be trimmed off (Figure 22a).

Muslin caps ready threaded with cotton are sometimes supplied. These are slipped over the bulb, and the thread is then pulled tight and tied (Figure 22b). The strands should be long enough to reach two or three inches below the lowest part of the bulb, in order that their lower ends can be immersed in the water vessel, but not long enough to hang in a bight, or water will drip from the wick at the lowest point of the curve until the reservoir is emptied.

24

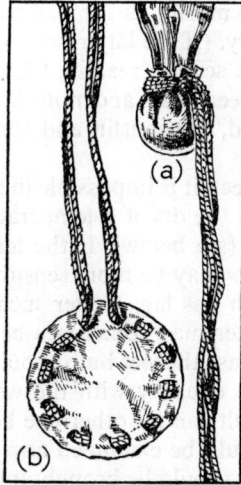

FIGURE 22. (*a*) Wet bulb with ordinary muslin and wick   (*b*) Muslin cap

Precautions necessary in taking wet-bulb observations. To get correct readings the muslin must be damp, but not dripping. If it is too wet, the reading of the thermometer will be too high. If it is not wet enough, the reading will again be too high. The former defect may be cured by cutting down the number of threads supplying moisture to the bulb. Take care, however, that this remedy does not make the muslin too dry.

It is important that the water should be pure. Ordinary water contains substances in solution and, if such water is used, as it evaporates it deposits these substances on the thread and muslin; the free flow of water to the muslin and its evaporation therefrom are checked, and the thermometer may read too high. Moreover, the rate of evaporation from impure water may differ appreciably from that for pure water. It is therefore desirable that distilled water should be used. This may be available from the ship's radio office, but is liable to become contaminated with acid in the course of a voyage. If, therefore, sufficient distilled water can be collected from the ship's radio office at the commencement of a voyage, this should be used. If distilled water is not available, condenser water from the engine-room may be used.

The muslin should be changed at least once a week and more often if it be dirty or contaminated by salt spray. The presence of salt in the water will cause the thermometer to read too high and, if any spray has reached the instrument, the muslin and wick should be replaced by new ones. It is advisable to do this in any case after bad weather. If it is found that an encrustation of lime or other impurity has formed on the thermometer bulb, this should be scraped off. A note should be made in the 'Remarks' column of the meteorological logbook whenever the muslin is changed.

After the muslin has been changed, some time must be allowed to elapse before observations are resumed. This is to ensure that the proper degree of wetting has been achieved and that the thermometer and wetted muslin have attained the properly balanced temperature.

Wet-bulb temperature higher than dry-bulb. If the reading of the wet-bulb thermometer is above that of the dry-bulb, first make sure that the readings were

D   25

correct. Then ensure that the muslin and thread are moist but not too wet and that the dry bulb is indeed dry. (If the latter has to be wiped, allow it to cool to the air temperature before a second reading.) If no fault is found, book the temperatures as they have been read and note in the 'Remarks' column that the reading has been checked, the muslin and thread examined, and that the ventilation is adequate.

Except as a result of a defect, it is impossible in normal circumstances for the wet bulb to read higher than the dry if a temperature is steady, and if the wet bulb is above freezing point (see below). If the temperature is changing, however, one of the thermometers may be more sensitive than the other and follow the temperature changes with less lag. Under such circumstances it is possible that the wet-bulb thermometer may sometimes be found to be reading higher than the dry bulb. In such a case the wet bulb should be taken as correct and the dry-bulb reading adjusted to equality with the wet bulb. If this phenomenon occurs frequently and the fault cannot otherwise be traced, it may lie in one of the thermometers. These should be examined and if there is nothing obviously wrong the spare thermometer should be brought into use to replace the first one, and then (if necessary) the other thermometer, till satisfactory observations are again obtained.

Wet-bulb readings during frost. During frost, when the muslin is thinly coated with ice, the readings are still valid because evaporation takes place from a surface of ice as freely as from one of water. If the muslin is dry it must be given an ice coating by wetting it slightly with ice-cold water, using a camel-hair brush or by other means. The water will usually take 10 to 15 minutes to freeze. Excess of water must not be used as it takes much longer to freeze and will also not give accurate readings. After the wetting of the muslin, the temperature generally remains steady at 0°C (32°F) until all the water has been converted to ice. It then begins to fall gradually to the true ice-bulb reading. No reading must be recorded until the temperature of the ice bulb has fallen below that of the dry bulb and remains steady. Dry windy weather may cause the ice to evaporate completely before the time of the next reading, in which case the procedure of wetting the bulb must be gone through again. The original coating of ice will give satisfactory results as long as it lasts.

It must be pointed out that supercooled water may exist on the wet bulb at temperatures well below freezing point and that, if this is not noticed by the observer, serious errors will occur. The freezing can be started by touching the wet bulb with a snow crystal, a pencil, or other object.

Measures of humidity. Dew-point and relative humidity can be obtained from the readings of the dry and wet bulb.

The *dew-point* is the temperature at which dew would begin to form on the bulb of the thermometer if the air were cooled down, the amount of water vapour in it remaining unchanged. Tables 15 to 18 give the dew-point for dry-bulb temperatures and depressions of the wet bulb. The depression of the wet bulb is the difference between the dry- and wet-bulb readings. The amount of this depression depends on the ventilation to which the wet-bulb thermometer is subjected and Tables 15 and 17 are to be used for observations in which the thermometers are exposed in the standard marine screen. Since the amount of evaporation from ice and water surfaces is not the same, lines are ruled in the tables to call attention to the fact that above the line evaporation is going on from a water surface while below the line it is going on from an ice surface. Intermediate figures must therefore be obtained by extrapolation.

26

In order that values of the wet-bulb depression of the necessary accuracy shall be available, it is especially desirable at low temperatures that the thermometers should be read to the tenth of a degree. This is because dew-point changes rapidly at low temperatures with changes in wet-bulb depression.

The *relative humidity* is the amount of water vapour actually present in the air, expressed as a percentage of the amount the air would contain at that temperature if it were saturated. Table 19 gives the relative humidity for dry-bulb temperature and depression of the wet bulb. In the British Meteorological Office a relative humidity of 95 per cent is taken as a guide in determining whether to report mist or haze (see tables on page 44).

ELECTRICAL RESISTANCE THERMOMETER Mk 2

The electrical resistance of platinum wire varies in a known way with changes of temperature. This characteristic is used in electrical resistance thermometers. Thin platinum wire in the form of a helix is enclosed inside a highly conductive ceramic former which is further encased in a close-fitting stainless-steel tubular sheath for protection. A plug of epoxy resin is placed in the head of the thermometer to prevent ingress of moisture to the platinum element. Electrical leads are brought through the head of the thermometer for connection to a suitable device to measure the change in resistance.

The Meteorological Office Resistance Thermometer Element Mk 2 has a diameter of 6 mm and a stem length of approximately 100 mm with a sensing length of 25 mm at the lower end of the stem. The thermometer is of the four-lead type, two of which are used for lead-compensation purposes.

When used as a wet-bulb thermometer a special tight-fitting tubular wick is fitted. Although the sensing part of the thermometer is located in the lower part, the full length of the stainless-steel stem conducts heat and it is essential that the wick is pushed up as far as possible, completely covering the stem, to ensure accurate wet-bulb readings.

The thermometers are generally used connected to the Meteorological Office Electrical Thermometer Indicator Mk 5, which is to be replaced by a digital readout thermometer indicator at a later date.

Electrical Thermometer Indicator Mk 5. Basically the indicator consists of an electrical a.c. balanced bridge circuit which is used to compare resistances. For temperature measurements the resistance of a remote Electrical Resistance Thermometer is compared with that of a precision resistor, known as the reference resistor. The difference is measured by means of a precision potentiometer and a temperature reading is taken from a suitably calibrated dial integral with it. The dial is calibrated from + 50 to − 50°C. Compensating leads are used to balance the resistance of the leads connecting the remote thermometer to the indicator.

The indicator and resistance thermometers will normally be installed by Meteorological Office technicians. If temperature is to be measured accurately the variable resistors in the electrical bridge must be carefully set to match individual thermometers and leads, and this can only be carried out with the correct test equipment. The user must not reset any variable resistors or change any thermometer or lead.

An illustration of the Electrical Thermometer Indicator Mk 5 is shown in Figure 9. Switch (a) is the on/off switch, switch (b) selects the thermometer to be used, with positions 1 to 6 for thermometers on long leads and 7 to 8 for thermometers on short leads. The electrical bridge inside the indicator is balanced by turning knob (c) and the balance indicated by the neon lamps (d). The temperature is read directly in degrees Celsius from the clock-type dial integral with knob (c). Switch (e) is used when checking the accuracy of the electrical bridge and for setting up the brightness of the neon balance indicator; it is normally left in the 'USE' position.

Operating the indicator:

(1) Use switch (b) to select the required thermometer.

(2) Make sure that switch (e) is in the 'USE' position.

(3) Press switch (a) and allow 10 seconds for the bridge to warm up.

(4) If the bridge is unbalanced one neon indicator will be on and the other off. If the right-hand neon indicator is off turn knob (c) clockwise, if the left-hand neon is off turn knob (c) anti-clockwise; the bridge is balanced when both neons are on.

(5) The temperature is read directly in degrees Celsius from the dial integral with knob (c). The small hand makes one revolution for 100°C (– 50 to + 50°C) and the large hand makes one revolution for 10°C. Each numbered division for the small hand is therefore 10° and for the large hand 1°. If the small hand is to the right of zero the temperature is positive and must be read clockwise from the black figures. If the small hand is to the left of zero the temperature is negative and must be read anti-clockwise from the red figures. Although it is possible to read the tens and units figures from the small hand and the units and tenths figures from the larger hand, it is, in practice, better to read the tens figure from the small hand and the units and tenths figures from the larger hand.

THE MARINE SCREEN

Exposure of thermometers. No matter how accurate a thermometer may be, it can do no more than indicate its own temperature. It is therefore essential that the thermometer is in correct contact with the medium whose temperature it is to measure, in order that it may 'share' its true temperature, and that it is protected from any extraneous source of heat. When measuring the air temperature, particular problems arise on this score, in that the thermometer must be shielded from the heat radiated by the sun, the sea and from the ship itself, yet at the same time the air, which is itself transparent to such radiated heat, must be allowed to flow freely past the thermometer. Even when not in use no normal meteorological thermometer, for whatever purpose it is supplied, must ever be exposed to full sunlight for more than a moment or two, and when stored, should be kept in its box.

A thermometer screen is used to shield the thermometer from external radiation, yet allowing an adequate flow of air.

Design of thermometer screen. There are several acceptable designs of thermometer screens, although only one is regarded as the Meteorological Office standard marine type. The essential features of any such screen, which at present is made of wood, are that the vertical walls are composed of louvres or 'jalousies', constructed so that no direct radiation can reach the thermometers but allowing relatively free air flow to reach the thermometer, while in addition vertical ventilation is permitted through the slotted floor and through holes in an inner roof. Hence, should the air within the screen become warmer or colder than its surroundings, it may rise or sink, and be replaced by outside air of the correct temperature. Such screens are painted white as a further precaution against radiation. Screens should be repainted when necessary and a watch should be kept for possible rotting of the woodwork, particularly in the lower corners. Marine screens will normally contain two thermometers, the dry bulb and the wet bulb.

Position of marine screen. The screen should be placed in the open air and, for convenience in reading the thermometers, about 1·5 metres above the deck. It may be exposed in sun or shade, preferably slung from an awning spar or ridge rope, so as to have an unimpeded circulation of air flowing through it. It should be out of the way of unauthorized persons; it must not be exposed to suddenly varying conditions due to causes within the ship, such as draughts of air from boilers, engine-room, etc. The lighting at night should be so arranged that it cannot affect the temperature of the thermometers. By day or by night the light should come from behind or from the side of the observer. The thermometer screen is usually placed on the bridge.

The position of the thermometer screen requires great attention. It cannot be too strongly emphasized that the temperature of the free air is required, not of that affected by heat from the ship. The most suitable location is where the air will come direct on to the screen from the sea before passing over any part of the ship. The ship is a source of local heat; radiation takes place from the hull and from sunny decks, deck houses, etc., especially in the tropics. Radiation of heat, or warm draughts of air, may be felt from galleys, engine and boiler rooms, stokehold and funnel. The thermometer screen should be as far as possible removed from all such sources of local heating which will tend to cause false air temperatures, particularly on days when the relative wind is light. The choice of the bridge will avoid some of these sources of heating.

The position of the screen may need changing with shifts of wind or alterations of course. When not in use it may be stowed as most convenient.

Setting up the thermometers. Sheathed thermometers are held in place by means of two clips. The thermometers should not be allowed to touch the floor of the screen but if they slip down through the clips a rubber grummet (obtainable from Port Meteorological Officers) should be placed over the thermometer, resting above the top or bottom clip. The plastic water bottle is held in place with a plastic or metal clip.

ASPIRATED AND WHIRLING PSYCHROMETERS

The system of measuring humidity by means of dry- and wet-bulb thermometers contained in a louvred thermometer screen assumes that the design of the screen controls the internal air flow between limits usually taken as 2–4 knots. Although in the open air the assumption appears reasonably correct, there will be occasions when greater accuracy is needed or, for temporary observations, a thermometer screen cannot be erected. Moreover the assumption will rarely, if ever, be true if temperatures and humidities are to be measured in confined spaces, for example within the hold of a ship.

The difficulty is overcome by artificially ventilating the thermometers at a controlled rate. Such thermometers are said to be 'aspirated'. Aspiration is performed by a fan which, operated by electric or clockwork motor, or even by hand (at a controlled speed) will draw air over the thermometers at a known rate. Aspirated thermometer systems include their own shield against direct radiation. An even more simple system ensures adequate ventilation by whirling or rotating the thermometers by hand at a controlled rate, the thermometers being mounted in a suitable holder to permit this. Such hand-held psychrometers are normally provided with no precautions against radiation whatsoever and must therefore be used only in the shade. This is also desirable in the case of the mechanically aspirated psychrometer.

As the rate of ventilation produced by aspirated and whirled psychrometers differs from that assumed to prevail in the static thermometer screen, different hygrometric tables must be employed, and the greatest care must be taken to ensure that the tables appropriate to the method are in fact used. Tables 16 and 18, at the end of this book, are those correctly employed with aspirated and whirling psychrometers. As with Tables 15 and 17, lines are ruled to draw attention to the fact that above the line evaporation is taking place from a water surface, while below the line it is occurring from an ice surface. Interpolation of readings must therefore not be made between figures on different sides of the line.

THE APPLICATION OF HYGROMETRIC OBSERVATIONS IN THE CARE AND PROPER VENTILATION OF CARGO

Sweating, or the deposition of moisture, is a frequent cause of damage, both to cargo and to the internal structure of a ship, and it is desirable to keep a record, not only of the temperature and humidity of the outside air through which the ship is passing, but also of the temperature and humidity of the air in each hold, as far as this is practicable. Although deductions from such data will vary according to the nature of the cargo and the construction of the ship, experience of these observations should help the seaman to judge whether, at any particular time, his cargo and the structure of his ship are in danger of damage by moisture and whether conditions are likely to be improved, or the reverse, by ventilation.

SEA TEMPERATURE

The routine meteorological requirement is for observation of sea-water temperature taken from near or just below the surface. The precise depth is not specified but any one of several methods is regarded as adequate. These methods are:

(a) by obtaining a sample by bucket;
(b) by thermometer immersed in the sea or in proximity to the sea;
(c) by engine-room intake temperature.

Bucket method. From a slow-moving ship having a bridge height of up to about 10 metres it is comparatively easy to draw a sample of sea water on board by almost any form of bucket strong enough to withstand the water pressure while being towed. A thermometer may then be inserted and the water temperature measured. Small buckets made of double-skinned canvas or rubber are very suitable for this purpose. Single-skinned canvas buckets are not suitable because any evaporation from the sides of the bucket would lower the temperature of the water sample.

The problem of getting a sea-water sample with a bucket becomes increasingly difficult as ships' size, speed and height of bridge are increased. Canvas buckets are so light that they would obviously be unsuitable for a fast ship from a high bridge. Even if not torn away on entry into the sea, little water would remain by the time it had been drawn up to deck level and the bucket's life would be very short. A smaller and somewhat heavier bucket made of rubber reinforced by canvas is now supplied to all British Voluntary Observing Ships. This bucket is little more than a closed length of rubber hose and it is suitable for taking sea temperatures in almost any ship, but a complete solution of successfully using a bucket regardless of the size and speed of ship has yet to be found. Extensive trials with this rubber bucket have shown that the temperature of the water sample changes very slowly after it has been hove on deck.

The small rubber buckets described above were originally designed to contain a thermometer which was lowered and immersed in the sea with the bucket itself. A high rate of thermometer breakage was experienced and the policy now is to immerse the thermometer into the sample of sea water when the bucket is drawn up on deck. There is in fact little disadvantage in this: whether the thermometer is immersed in the sea or inserted later, it will do no more than measure the temperature of the sample at the moment of observing.

Whichever type of bucket is used, it should be swung as far out as possible to avoid the shallow layers of water close to the hull which have been warmed by the ship itself. Probably the best way of getting the water sample is to use the bucket as though one were taking a cast of the hand lead. On entering the water the bucket should submerge quickly and cleanly. If drawn along the surface, a fault to which some designs are particularly prone, it will be filled with spray, possessing some temperature intermediate to that of the sea and that of the air.

On being withdrawn, a thermometer should be inserted into the sample immediately. This should be done in the shade; direct sunlight, in addition to its direct effect upon the thermometer, can warm the sea-water sample very quickly.

Individual thermometers are calibrated either for complete immersion into the medium whose temperature is to be measured, or for contact through the thermometer bulb alone (e.g. clinical thermometers). Meteorological thermometers are invariably of the former class and, if not large, would give rise to unacceptable errors when the air/sea temperature differences are large. In consequence the whole thermometer should be covered by the sea water without touching either the sides or bottom of the bucket. Devices which hold the thermometer within the bucket may be available, but otherwise it should be held at the extreme end by finger tip, without actually letting the fingers (which are a source of heat) enter the sample. With the large canvas bucket the thermometer should be moved with a slow stirring action. After immersion for about one minute the thermometer should be withdrawn just sufficiently to allow the scale to be read, the bulb and as much of the stem as possible being left immersed.

Almost any meteorological thermometer may be used for this purpose. Where special support frames called sea protectors (Figure 7b) are employed for use with the canvas bucket (Figure 10) the thermometer must be of suitable dimensions, but otherwise it is convenient to use the type common to those employed for dry- and wet-bulb observation.

After use, the thermometer should be dried and returned to its box for careful storage with the bulb end downwards.

Distant-reading thermometers. There would obviously be many advantages in measuring temperature by means of a distant-reading instrument while the thermometer bulb was actually immersed in the sea. In its most simple form such a device would be lowered by cable alongside the ship and readings taken in-board while it was towed. There are, however, certain difficulties in such a method. It is difficult to control the depth of such a device or even ensure that it enters the water at all and does not merely skip along the surface. The strain of towing upon the cable can also be a cause of error in the electrical measurements, while a freely towed device could damage itself against the side of the ship. Such devices are still being developed experimentally but no standard generally usable instrument is yet available.

Some ships, mainly of foreign origin, employ a submerged electrical thermo-meter attached to the hull of the ship, indicating its observations by cable to the bridge. This is most convenient, but the thermometer, even though partially protected, is vulnerable to damage by flotsam or when in dock and the hull itself must be pierced to allow entry of the cable.

A method evolved by the Meteorological Office and installed in a few modern ships is somewhat similar but places the thermometer inside the hull, measuring the sea temperature by conduction through the ship's side plating, the principle being that steel is such a good conductor that it transmits the temperature of the surrounding sea water. The thermometer, which is in the form of a small, thin, printed electrical resistance circuit little bigger than a postage stamp, is fixed to the inside of the hull in a forward and unheated compartment of the ship at a point a metre or so below the normal water-line. The system which requires the whole plate to change temperature with that of the sea, has a long time lag, and is thus unaffected by short-period roll or pitch, but would be invalidated if the position of the thermometer were raised above sea level by change in loading. The system demands cabling to the place where temperatures are to be read, normally the bridge, and installation is therefore somewhat costly.

Engine-room intake temperatures. The temperature of the engine-room intake water can be taken as a measure of sea-water temperature either by thermometer or by thermograph. To an extent dependent on the individual ship, the accuracy will be questionable although the method is very convenient and may well be the only one possible (in the absence of the hull thermometer described above) when the bucket method cannot be used because of rough seas, too great a ship speed or a bridge too high above the water. The errors arise from the varying depth from which the water is drawn as the ship rolls or pitches and the risk of pre-heating as the water passes through pipes at or close to engine-room temperature or through oil and water tanks on the inside of the hull. A sample of the intake water may be drawn off by tap, the subsequent procedure being that described in the bucket method above, or the temperature measured by a thermometer installed within the intake pipe. In the latter case the thermometer will usually be inserted in a pocket formed within the pipe, and the main prob-lem which then arises is of assuring good thermal conductivity. One of the disadvantages of using intake temperatures is the liability to error due to having to telephone the thermometer reading from the engine-room to the

32

bridge. Obviously it would be possible to have a distant-reading arrangement so that the intake temperature could be read on the bridge but this would be an expensive arrangement.

THE MEASUREMENT OF SOLAR RADIATION

Solar radiation (the electromagnetic radiation from the sun) is the prime driving force of all atmospheric disturbances so it is important to measure the energy from this source received at different parts of the globe, both on land and at sea. This radiation consists of a large range of wave-lengths, but three broad categories can be recognized:

(a) Wave-lengths too short for the eye to see, known as ultra-violet radiation.
(b) Wave-lengths to which the eye is sensitive, known as light.
(c) Wave-lengths too long for the eye to see, known as infra-red radiation.

At sea the measurement is usually of the total energy in all these wave-lengths which falls on a unit horizontal area. About half the energy measured is, in fact, light and the remainder is either ultra-violet or infra-red radiation. In the absence of cloud or an atmosphere all the solar radiation would, of course, come only from the direction of the sun and the remainder of the sky would be dark, but in natural conditions a considerable portion of the energy is received as diffuse or scattered radiation. When the sky is overcast all the radiation received is diffuse.

The instrument in general use in the Meteorological Office for measuring solar radiation is a Moll-Gorczynski solarimeter or pyranometer (see Figure 15). It is a horizontal thermopile (with the surface blackened to increase absorption) with alternate thin strips of manganin and constantan, one set of junctions being along the centre line of the surface (the active junctions) while the remaining junctions are in good thermal contact with the relatively massive supporting posts, which are insulated electrically but not thermally from the base plate. When radiation falls on this thermopile the temperature of the centre is raised and a small voltage (some millivolts) is produced between the leads, this voltage being recorded as described below. The thermopile is shielded by two glass domes and the outer one needs to be wiped clean whenever necessary—particularly of spray deposits at sea; the interior of the instrument is kept dry by silica gel (in a screw-off holder) which must be renewed when necessary. A white horizontal guard plate, removed for the photograph (Figure 16), is mounted level with the thermopile surface to help protect the outside of the instrument from direct radiation.

All the British Ocean Weather Ships have these solarimeters in a special gimbal mounting, so that the thermopile remains horizontal, and they must naturally be fitted in as unobstructed a position as possible—normally high on the bridge. The recording is usually on a chart of a self-balancing potentiometer, but the Meteorological Office Data Logging Equipment (Modle), now in use at land stations and aboard the British Ocean Weather Ships, produces a chart, a digital display and punched paper tape (later to be magnetic tape), the tapes being processed by the Meteorological Office computer to yield the required mean values.

Hydrographic survey ships of the Royal Navy, British research ships and certain British Selected Ships have also been supplied with these special mount-

33

ings; here the recording is mostly by the chart display of a self-balancing potentiometer, but increasing use is being made of a direct magnetic-tape recorder, which can be processed by the computer.

Those ships having solarimeters (pyranometers) receive special installation and maintenance instructions and there is a regular exchange of recalibrated instruments arranged by the National Radiation Centre of the Meteorological Office.

CHAPTER 3

General Meteorological Instruments

Instruments have been developed for land stations for the measurement of almost every meteorological element, with the notable exception of cloud type and amount. These, being purely subjective observations, can only be assessed by human interpretation. Unfortunately very few of the instruments designed for land stations are capable of operating with reasonable satisfaction on board ship where special problems arise, not only from the motion of the vessel, its forward movement and exposure to saline spray, but also because of the disturbances created by the ship itself in the surrounding atmosphere.

In consequence, methods suitable for land stations would be at best merely an unnecessary expense if used at sea, and at worst would soon fail to operate or give misleading results. For example, a hair hygrograph, an instrument giving a continuous recording of the relative humidity of the air, relies upon the change of length of a bundle of human hairs for its operation. In addition to the adverse effect of ship's movement upon the very light and delicate pen movement, the existence of salt upon the hairs would seriously affect their readings.

Pressure, temperature and, to a large degree, humidity have been covered in the preceding chapters. For general interest the standard instruments for measuring the other normally observed elements are given below, without regard as to whether they might, in special circumstances, be adapted for ship use.

WIND SPEED AND DIRECTION

Anemometer. This consists of cups rotatable about a vertical shaft or a propeller rotating about a horizontal shaft. When driven by the wind at a speed proportional to wind force, the rotating shaft drives an electrical generator whose output is itself proportional to the speed of rotation. A voltmeter may thus be calibrated as wind speed. Anemometers are not normally used aboard merchant ships because of the difficulty of finding a suitable site and also because of expense. The British Weather Ships carry two anemometers on a yardarm, one each side of the main mast at a height of 20 metres above the water, which seems to be the site furthest from eddying effects. But even here estimates are made regularly of wind force and direction from the appearance of the sea as a check on the instruments. (See Figures 13 and 14.)

Wind vane. The wind direction may be directly observed by the position of the wind vane or remotely read at ground (or deck) level by an electrical direction transmitter known as a Magslip (for mains power) or a Desynn (battery operation).

RAINFALL

Rain-gauge. Rainfall is collected over a known horizontal area by means of a funnel placed over a metal can or glass bottle. The water so collected may be stored for measurement as a total quantity over an interval of 12 or 24 hours, or be made to operate one of various recording devices.

35

CLOUD HEIGHT

Cloud-height balloons. These are small balloons filled with hydrogen to give them a known rate of ascent. When released from the ground they are timed with a stopwatch until seen to enter the cloud base. From the rate of ascent and the time taken the cloud height is then computed.

Cloud searchlight. This method is only available at night. A small searchlight is projected vertically on to the cloud base and observed from a known distance, usually about 300 metres. From the point of observation, the apparent angle of the bright spot where the searchlight falls on to the cloud is observed by alidade, theodolite or sextant and the height can thus be calculated.

Cloud-base recorder. This is an automatic device in which the same principles are employed as for the cloud searchlight, in that the angle at which a projected beam of light meets the cloud is measured over a known baseline. In this automatic instrument, the light is 'modulated', i.e. is interrupted at a known rate, and a photocell, replacing the human eye, detects only this modulated light even in the presence of daylight. The cloud-base recorder can thus operate day or night and will provide a continuous record of cloud height from its own operation of the angle of sight measured by the photocell system.

UPPER WINDS

Pilot-balloon ascent. Small hydrogen-filled balloons, whose rate of ascent is predetermined, are followed by optical theodolite. From the observed bearings the horizontal movement of the balloon, and hence the wind, can be computed for each layer of the atmosphere through which the balloon passes. This method is rarely used in routine practice.

Wind-finding radar. This method is similar to the pilot-balloon ascent but the balloon is followed by radar. In consequence, the upper winds may be computed even when the balloon is invisible in cloud or fog. The balloon carries a 'radar reflector' to provide proper radar response. The radar is specially designed to give precision bearings of azimuth and elevation and accurate ranges.

UPPER-AIR TEMPERATURE AND HUMIDITY

Radiosonde. A device carried by hydrogen-filled balloons to make frequently repeated observations of pressure, temperature and humidity and to signal these observations to the ground by radio (see Figure 13). The radiosonde is frequently carried aloft by the same balloon used for radar-winds. Special ground equipment and staff training is usually required to receive and decode the radiosonde signals. Consequently observations of this kind are not made from British merchant ships although they are made regularly aboard weather ships.

VISIBILITY

Visibility recorder or transmissometer. Light from a lamp of known output is transmitted over a prescribed distance parallel to the ground. It is then received by a calibrated photocell. The degree by which the light has been attenuated by mist or fog is thus observable. An alternative system, employing the increased amount of light scattered backwards towards the light when fog exists, is already in use by Trinity House on lighthouses and light-vessels.

Part II Non-instrumental Observations

Introduction. Non-instrumental observations are very important and, being estimates, they are dependent upon the personal judgment of the observer. This judgment is the product of training and experience at sea, together with practice in making the observations. To acquire a technique of observation, adherence to the official instructions is essential. The aim of these instructions is not only to outline a satisfactory method of making observations, but to impose a standard procedure such that two observers, despite differences in training, will make approximately the same observation in similar circumstances. The assumption that observations are comparable, or made according to the same procedure, is the basis of synoptic meteorology or of a study of climate.

Observations from ships are of special importance to the forecaster not only because they enable him to complete his charts over the oceans, but also because weather sequences at sea are simpler than those on land. They are therefore more characteristic of the air masses and hence more useful in the air-mass analysis that precedes the preparation of forecasts. Numerous instances occur in which the presence or absence of adequate ship reports has made all the difference between good and bad weather forecasts. An observer should never forget that his individual effort, his particular observations, may supply just the information required to resolve a forecasting problem hundreds or thousands of miles away.

The making of meteorological observations at sea is attended by many difficulties that are unknown to the observer on shore. It is in overcoming them that the experience and training of the mariner are important. These difficulties largely result from the movement of the ship and the absence of landmarks.

CHAPTER 4

Wind, Weather and Visibility

Wind force and direction. Wind force is expressed numerically on a scale from 0 to 12.* This scale, which originally defined the wind force in terms of the canvas carried by a frigate, was devised by Captain, afterwards Admiral, Sir Francis Beaufort in the year 1808 for use in vessels of the Royal Navy. Since Admiral Beaufort's time, however, so many changes had taken place in the build, rig, and tonnage of sea-going vessels that in 1874 Beaufort's scale was adapted to the full-rigged ship with double topsails of that period. With the passing of sail, this specification meant very little to those who had no experience in square-rigged ships, and the practice arose of judging wind force from the state of the sea surface. In 1939 the International Meteorological Organization agreed to the use of a sea criterion by which the wind force was judged from the appearance of the sea surface. This specification, brought into use in 1941, is shown on pages 38 and 39. Photographs showing the appearance of the sea corresponding to each Beaufort force are given between pages 40 and 41.

---

*See Note to Table 23.

BEAUFORT SCALE OF WIND FORCE

Beaufort scale number	Mean wind speed in knots	Limits of wind speed in knots	Descriptive terms	Sea criterion	Probable height of waves in metres*	Probable maximum height of waves in metres*
	Measured at a height of 10 metres above sea level					
0	00	Less than 1	Calm	Sea like a mirror.	—	—
1	02	1–3	Light air	Ripples with the appearance of scales are formed but without foam crests.	0·1	0·1
2	05	4–6	Light breeze	Small wavelets, still short but more pronounced; crests have a glassy appearance and do not break.	0·2	0·3
3	09	7–10	Gentle breeze	Large wavelets. Crests begin to break. Foam of glassy appearance. Perhaps scattered white horses.	0·6	1·0
4	13	11–16	Moderate breeze	Small waves, becoming longer; fairly frequent white horses.	1·0	1·5
5	19	17–21	Fresh breeze	Moderate waves, taking a more pronounced long form; many white horses are formed. (Chance of some spray.)	2·0	2·5
6	24	22–27	Strong breeze	Large waves begin to form: the white foam crests are more extensive everywhere. (Probably some spray.)	3·0	4·0
7	30	28–33	Near gale	Sea heaps up and white foam from breaking waves begins to be blown in streaks along the direction of the wind.	4·0	5·5
8	37	34–40	Gale	Moderately high waves of greater length; edges of crests begin to break into spindrift. The foam is blown in well-marked streaks along the direction of the wind.	5·5	7·5

BEAUFORT SCALE OF WIND FORCE—continued

Beaufort scale number	Mean wind speed in knots	Limits of wind speed in knots — Measured at a height of 10 metres above sea level	Descriptive terms	Sea criterion	Probable height of waves in metres*	Probable maximum height of waves in metres*
9	44	41–47	Strong gale	High waves. Dense streaks of foam along the direction of the wind. Crests of waves begin to topple, tumble and roll over. Spray may effect visibility.	7·0	10·0
10	52	48–55	Storm	Very high waves with long overhanging crests. The resulting foam in great patches is blown in dense white streaks along the direction of the wind. On the whole the surface of the sea takes a white appearance. The tumbling of the sea becomes heavy and shocklike. Visibility affected.	9·0	12·5
11	60	56–63	Violent storm	Exceptionally high waves. (Small and medium-sized ships might be for a time lost to view behind the waves.) The sea is completely covered with long white patches of foam lying along the direction of the wind. Everywhere the edges of the wave crests are blown into froth. Visibility affected.	11·5	16·0
12	—	64 and over	Hurricane	The air is filled with foam and spray. Sea completely white with driving spray; visibility very seriously affected.	14 or over	—

*These columns are added as a guide to show roughly what may be expected in the open sea, remote from land. In enclosed waters, or when near land with an off-shore wind, wave heights will be smaller and the waves steeper.

NOTES.—(a) It must be realized that it will be difficult at night to estimate wind force by the sea criterion.

(b) The lag effect between the wind getting up and the sea increasing should be borne in mind.

(c) Fetch, depth, swell, heavy rain and tide effects should be considered when estimating the wind force from the appearance of the sea.

In using this specification it is assumed that the observation is made in the open ocean and that the wind has been blowing long enough to raise the appropriate sea. The possibility of a lag between the wind getting up and the sea increasing must be considered. The appearance of the sea surface also depends on many other factors such as the fetch of the wind (i.e. distance from weather shore), the swell, the presence of tides, and whether or not precipitation is occurring. These effects should be allowed for before deciding the appropriate number on the scale. Experience is the only sure guide but the following remarks may be of some use:

(a) A discrepancy between wind and sea occurs frequently close inshore where winds of a local character are likely.

(b) An off-shore wind does not produce its appropriate sea close inshore but requires a certain fetch before its full effect is produced.

(c) Swell is the name given to waves, generally of considerable length, raised by winds at a considerable distance from the point of observation. Swell is not taken into account when estimating wind.

(d) Tides or strong currents affect the appearance of the sea surface, a wind against tide or current causing more 'lop'—a weather tide—and the wind in the same direction as a tide or current producing less disturbance of the sea surface—a lee tide.

(e) Precipitation, especially if heavy, produces a smoothing effect on the sea surface.

(f) There is evidence that the height of the sea disturbance caused by a wind of a particular force is affected by the difference between sea and air temperatures, the sea being the warmer medium. If this difference increases, there is an appreciable increase in the sea disturbance, and vice versa.

Beaufort force can be transformed approximately into wind speed by means of a table of equivalents included in the specification of Beaufort force given on pages 38 and 39.

The International Code (used for making meteorological reports by radio) makes provision for the reporting of wind speed in knots. The observer may derive this from the table of equivalents, taking the mid point of the range corresponding with the observed Beaufort force; or, better still, he may interpolate according to his own judgment. For example, if the wind is estimated to be over Beaufort 5 but not quite Beaufort 6, it might be reported as having a mean speed of 21 knots.

Wind direction is logged as the true, not the compass, direction and is given to the nearest ten degrees. The exposed position that a ship's standard compass usually occupies gives a clear all-round view and from it the observer takes a compass bearing, noting the tops of the waves, the ripples, the spray and the faint lines that generally show along the wind. It is usually best to look to windward in judging wind direction, but in some lights the direction is more evident when looking to leeward.

Meteorologists as well as seamen use the term 'veering' to indicate a change of wind in a clockwise direction and the term 'backing' to denote a change in an anticlockwise direction.

Estimation of wind force and direction can often be made in the same way at night but sometimes on very dark nights it is impossible to see the effect of

STATE OF SEA PHOTOGRAPHS, FOR ESTIMATING WIND SPEED
(SEE PAGE 37)

See footnotes on page 39

Photograph by R. R. Baxter (Crown Copyright)

FORCE 0

Wind speed less than 1 kn. (Sea like a mirror.)

Photograph by R. R. Baxter (Crown Copyright)

FORCE 1

Wind speed 1–3 kn; mean, 2 kn.
(Ripples with the appearance of scales are formed, but without foam crests.)

SEA PLATE I

E

Photograph by R. R. Baxter (Crown Copyright)

FORCE 2

Wind speed 4–6 kn; mean, 5 kn.
(Small wavelets, still short but more pronounced—crests have a glassy appearance and do not break.)

Photograph by R. Palmer

FORCE 3

Wind speed 7–10 kn; mean, 9 kn.
(Large wavelets. Crests begin to break. Foam of glassy appearance. Perhaps scattered white horses.)

SEA PLATE II

Photograph by P. J. Weaver

FORCE 4

Wind speed 11–16 kn; mean, 13 kn.
(Small waves, becoming longer; fairly frequent white horses.)

Photograph by R. R. Baxter (Crown Copyright)

FORCE 5

Wind speed 17–21 kn; mean, 19 kn.
(Moderate waves, taking a more pronounced long form; many white horses are formed.
Chance of some spray.)

SEA PLATE III

FORCE 6

Wind speed 22–27 kn; mean, 24 kn.
(Large waves begin to form; the white foam crests are more extensive everywhere. Probably some spray.)

FORCE 7

Wind speed 28–33 kn; mean, 30 kn.
(Sea heaps up and white foam from breaking waves begins to be blown in streaks along the direction of the wind.)

SEA PLATE IV

Photograph by R. R. Baxter (Crown Copyright)

FORCE 8

Wind speed 34–40 kn; mean, 37 kn.
(Moderately high waves of greater length; edges of crests begin to break into spindrift. The foam is blown in well-marked streaks along the direction of the wind.)

Photograph by O. R. Bates

FORCE 9

Wind Speed 41–47 kn; mean, 44 kn.
(High waves. Dense streaks of foam along the direction of the wind. Crests of waves begin to topple, tumble and roll over. Spray may affect visibility.)

SEA PLATE V

Force 10

(The upper and lower photographs illustrate the difference in appearance between seas viewed along the trough and at right angles to the trough respectively.)

Wind speed 48–55 kn; mean, 52 kn.
(Very high waves with long overhanging crests. The resulting foam, in great patches, is blown in dense white streaks along the direction of the wind. On the whole, the surface of the sea takes on a white appearance. The tumbling of the sea becomes heavy and shock-like. Visibility affected.)

Sea Plate VI

Photograph by Kevin O'Keeffe (Crown Copyright)

Photograph by Post Office (Crown Copyright)

FORCE 11

(The upper and lower photographs illustrate the difference in appearance between seas viewed along the trough and at right angles to the trough respectively.)

Wind speed 56–63 kn; mean, 60 kn.
[Exceptionally high waves. (Small and medium sized ships might be for a time lost to view behind the waves.) The sea is completely covered with long white patches of foam lying along the direction of the wind. Everywhere the edges of the wave crests are blown into froth. Visibility affected.]

SEA PLATE VII

Photograph by Post Office (Crown Copyright)

FORCE 12

Wind speed 64–71 kn; mean, 68 kn; the Beaufort Scale actually extends to Force 17 (up to 118 kn), but Force 12 is the highest which can be identified from the appearance of the sea. (The air is filled with foam and spray. Sea completely white with driving spray; visibility very seriously affected.)

SEA PLATE VIII

the lighter winds on the sea surface. In such cases the apparent or relative wind force and direction must be estimated by their effect, i.e. by the 'feel' upon the face or upon a moistened finger, or by the direction in which the smoke is blowing. Allowance must then be made for the ship's course and speed. In a fast ship considerable difference exists between the apparent and true wind directions. When the wind is astern and of the same velocity as the ship there is apparent calm on board the ship. In a calm a ship steaming 10 knots will have an apparent head wind of velocity 10 knots, but as soon as the wind blows from any direction out of the fore and aft line, the difference between the apparent and true directions will vary with each angle on the bow, and each force of the wind. The true wind may be obtained from the apparent wind by use of the parallelogram of velocities, or Table 22 as explained below. In Figure 23 if, for example, the ship is travelling along the line AB with speed 15 knots and the wind appears to be coming from the direction DA with speed 29 knots (Beaufort scale 7), the true direction of the wind is along CA and its speed 18 knots.

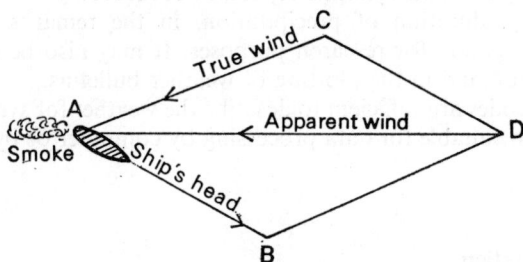

FIGURE 23. Wind, parallelogram of velocities

This result is easily obtained graphically by drawing the figure, making BA proportional to 15 and DA proportional to 29, and then measuring DB which is equal to CA, where ABDC is a parallelogram. The angle CAD, which is the same as BDA, is measured with a protractor and gives the difference between the true and apparent directions of the wind. Table 23 enables the conversion from apparent to true wind to be made by inspection.

In fast vessels the task of estimating accurately the true wind force and direction is no easy one and special care is required; this applies particularly to occasions when the wind is very light, and on dark nights.

Anemometers have as yet found only limited use at sea. the chief problem being to achieve a suitable exposure. The ship disturbs the airflow in its vicinity with the result that the wind measured by the instrument is not representative of the true airflow over the open sea. If a portable cup anemometer is used, the exposure may be varied at will and the best position chosen for any particular wind direction. The instrument measures 'apparent' wind speed. To determine the true value, the wind direction must first be estimated and then allowance made for the speed of the ship.

Wind force and direction, taken alone, do not completely specify the character of the wind. It is well known that on occasions the wind is particularly gusty, as in showery weather. Less frequently, definite squalls may occur. The difference between a gust and a squall is essentially one of time-scale, a gust being momentary, whereas a squall may last several minutes. It is

F   41

important when making the observations to note any unusual gustiness and the occurrence of squalls. When the latter occur it is of advantage if the time be noted together with any sudden change in wind direction. It is of interest to note that gusts have no appreciable effect in raising waves, whereas squalls may act for a sufficient length of time to raise a group of waves which tend to travel with the squall.

WEATHER

For the purposes of the meteorological logbook, the term 'weather' embraces those elements covered by the 'present weather' and 'past weather' codes, i.e. fog, precipitation, etc. (See Met. O. 509, *Ships' Code and Decode Book*.)

For a concise description of weather, Admiral Beaufort devised a system known as the Beaufort notation. Since 1958 this method has not been used to record weather at the synoptic hour but it is given below as it provides a handy way of amplifying the main synoptic report, or of recording the weather between observations, e.g. duration of precipitation, in the remarks column of the meteorological logbook for research purposes. It may also be found useful in the Deck Logbook and in the plotting of weather bulletins.

The present codes are sufficient to describe the weather for synoptic purposes, and they are also suitable for data processing by computer or by other machine methods.

The Beaufort notation

Weather	Beaufort letter	Weather	Beaufort letter
Blue sky (0–2/8 clouded) . .	b	Overcast sky. (The whole sky covered with unbroken cloud)	o
Sky partly clouded (3–5/8 clouded)	bc	Squally weather . . .	q
Cloudy (6–8/8 clouded) . .	c	Rain	r
Drizzle	d	Sleet (i.e. rain and snow together)	rs
Wet air (without precipitation) .	e	Snow	s
Fog	f	Thunder	t
Gale	g	Thunderstorm with rain or snow	tlr or tls
Hail	h	Ugly threatening sky . .	u
Precipitation in sight of ship or station	jp	Unusual visibility . . .	v
Line squall	kq	Dew	w
Storm of drifting snow . .	ks	Hoar frost	x
Sandstorm or duststorm . .	kz	Dry air	y
Lightning	l	Dust haze	z
Mist	m		

The system has been extended since Beaufort's day to provide indication of intensity and continuity. Capital letters are now used to indicate occasions when the phenomenon noted is intense. On the other hand, occasions of slight intensity are distinguished by adding a small suffix 'o'.

Thus R = Heavy rain.
r = Moderate rain.
r_0 = Slight rain.

The prefix 'i' indicates 'intermittent', thus:

    if = Fog patches.

    ir_0 = Intermittent slight rain.

The prefix 'p' indicates 'shower of', thus:

    pR = Shower of heavy rain.

    ps_0 = Shower of slight snow.

A solidus '/' is used in 'present weather' to distinguish present conditions from those in the past hour, thus:

    c/r_0 = Cloudy after slight rain in the past hour.

Continuity is indicated by repeating the letter, thus:

    rr = Continuous moderate rain.

The following are further examples of the use of Beaufort notation:

    cs_0s_0 = Cloudy with continuous slight snow.

    oid_0 = Overcast with intermittent slight drizzle.

    bif   = Blue sky with fog patches.

    cqprh = Cloudy with squalls and shower of moderate rain and hail.

    crrm = Cloudy with continuous moderate rain, and mist.

In past weather the letters are used in the same way but their order from left to right indicates sequence in time.

Thus 'b, bc, cpr' indicates cloudless conditions, becoming partly cloudy, followed by cloudy conditions with shower(s) of rain.

Precipitation. A distinction is drawn in the present and past weather codes between rain, drizzle and showers. Showers are of short duration and the fair periods between them are characterized by clearances of the sky. Showers fall from clouds having great vertical extent and usually isolated. They do not often last more than half an hour. Showers are characteristic of an unstable polar air mass, usually flowing in the rear of a depression, but they are by no means confined to this situation.

Rain and drizzle fall from overcast or nearly overcast skies. The distinction between rain and drizzle depends not on the amount of the precipitation but on the size of the drops. Drizzle is 'precipitation in which the drops are very small' (diameter less than 0·5 mm). Slight rain, on the other hand, is precipitation in which the drops are of appreciable size (they may even be large drops), but are relatively few in number. Observers should decide from the size of the drops whether the precipitation is drizzle or rain, and from the combined effect of the number and size of the drops whether the precipitation is slight, moderate, or heavy. The description 'heavy' is relatively rare in temperate latitudes.

Precipitation is defined as intermittent if it has been discontinuous during the preceding hour, without presenting the character of a shower. Observers should cultivate the practice of recording the times of onset and cessation of precipitation.

Fog, mist and haze. Fog, mist and haze have in the past been used, rather loosely, to describe decreasing degrees of obscurity in the atmosphere. Modern practice reserves the description 'haze' for occasions when the obscurity is caused by solid particles such as dust or sea salt. Fog and mist are akin in that they are both composed of minute water drops and may thus be distinguished from haze. In practice the distinction is usually made by means of the dry- and wet-bulb readings. The following table gives the approximate criterion for the reporting of mist and haze at various temperatures. Intermediate values may be obtained by interpolation. If the depression of the wet bulb is more than about that shown in the relevant column B, haze should be reported. If the depression is less, the obscurity should be reported as mist. (A relative humidity of 95 per cent is used by the British Meteorological Office as a guide to the dividing line between mist and haze.)

DEPRESSION OF THE WET BULB CORRESPONDING TO A RELATIVE HUMIDITY OF 95%

Column A	Columns B		Column A	Columns B	
Dry Bulb °C	Depression, °C		Dry Bulb °F	Depression, °F	
	Stevenson screen	Aspirated psychrometer		Stevenson screen	Aspirated psychrometer
40	0·8	0·8	100	1·3	1·4
35	0·7	0·7	90	1·2	1·3
30	0·7	0·7	80	1·2	1·2
25	0·6	0·6	70	1·0	1·0
20	0·5	0·6	60	0·8	0·9
15	0·5	0·5	50	0·7	0·7
10	0·4	0·4	40	0·6	0·6
5	0·3	0·3	32	0·4	0·5
0	0·3	0·3			

The further distinction between mist and fog is only one of degree and is arbitrarily assigned. When the visibility is reduced to less than 1 km or 0·54 n. mile the obscurity is described as fog; when greater than 1 km it is known as mist.

Visibility. Although the use of such terms as fog, mist and haze is suitable for a general indication of the state of visibility in the ww code or in the text of a ship's logbook, a more precise method is needed in weather messages to indicate to the meteorologist the degree of obscurity of the atmosphere, irrespective of the reason that causes it. On land, observations are made of a number of selected objects at fixed distances, the distances increasing roughly in such a way that each distance is nearly double the next smaller distance. The determination of the most distant object of the series which is visible on any given occasion constitutes the observation of visibility. At sea such a detailed determination of visibility is not usually possible, but in making estimates of visibility a coarser scale is used, as shown below.

VISIBILITY SCALE FOR USE AT SEA

Code figure		Code figure	
90	Less than 50 m	95	1·1 n. mile or 2 km
91	50 m	96	2·2 n. mile or 4 km
92	0·11 n. mile or 200 m	97	5·4 n. mile or 10 km
93	0·27 n. mile or 500 m	98	10·8 n. mile or 20 km
94	0·54 n. mile or 1 km	99	27·0 n. mile or 50 km

Note 1. If the distance of visibility is between two of the distances given in the table, the code figure for the shorter distance is reported.
Note 2. The prefix '9' before each of the scale numbers appears here because this table is part of a code for reporting visibility in two figures by radio (see *Ships' Code and Decode Book* (Met. O. 509) or *Admiralty List of Radio Signals*, Vol. 3).

In a long vessel the determination of the lowest numbers offers no difficulty as objects at known distances may be used. Visibility numbers in the middle range indicate conditions of obscurity such that the visibility is greater than the length of the ship but is not sufficient to allow full speed to be maintained. The only means of obtaining observations for the higher numbers of the scale are as follows. When coasting and when fixes can be obtained, the distance of points when first sighted, or last seen may be measured, from the chart. In the open sea, when other ships are sighted, visibility may be estimated by noting the radar range when the vessel is first sighted visually and again when it disappears from view. It is customary to use the horizon to estimate visibility numbers in the higher range although this cannot be relied upon. There are cases of abnormal refraction when the visible horizon may be very misleading as a means of judging distances, particularly when the height of the eye is great, as in the case of an observer on the bridge of a large liner.

The estimation of visibility at night is very difficult. What the meteorologist is interested in knowing is the degree of transparency of the atmosphere. But the distance seen at night depends on the amount of illumination; and the distance at which a light is seen depends on its intensity or candle-power. If there is no obvious change in meteorological conditions, the visibility just after dark will be the same as that recorded just before dark irrespective of the fact that one may not be able to see as far. A deterioration in visibility can sometimes be detected afterwards and the visibility figure adjusted accordingly. In doing this, care must be taken not to confuse the effect of a decrease in illumination, as for example when the moon sets, with a genuine decrease in visibility. The presence of a 'loom' around the vessel's navigation lights is frequently a guide to deteriorating visibility.

CHAPTER 5

Clouds and Cloud Height by Estimation

A normal observation of cloud at sea involves:

(a) The identification of the cloud types present.
(b) An estimation of the height of the base of the lowest cloud in the sky.
(c) An estimation of the amount of all cloud of type C_L (or C_M, if no C_L).

The fundamental distinction in structure, which has great significance for forecasting, is between 'layer' or 'sheet' clouds, and 'heap' clouds, i.e. clouds with marked vertical development. Examples of the latter are cumulus, sometimes known as the 'wool pack' or 'cauliflower' cloud, and cumulonimbus, the 'thundercloud' or 'anvil' cloud. In the further classification of sheet or layer clouds the consideration of height is taken into account, but the classification is not strictly one of height so much as of appearance. The main classification is into ten types as follows:

Sheet clouds   *Approximate limits (see also page 52)*

Cirrus	(Ci)	
Cirrocumulus	(Cc)	Base above 18 000 feet (5500 m)
Cirrostratus	(Cs)	
Altocumulus	(Ac)	6500 to 18 000 feet (2000 to 5500 m)
Altostratus	(As)	
Nimbostratus*	(Ns)	
Stratocumulus	(Sc)	
Stratus	(St)	
		Base below 6500 feet (2000 m)

Heap clouds (with vertical development)
Cumulus   (Cu)
Cumulonimbus   (Cb)

Descriptions of the different types are given below.†

Cirrus (Ci). Detached clouds of delicate and fibrous appearance, without shading, generally white in colour, often of a silky appearance. Cirrus appears in the most varied forms, such as isolated tufts, lines drawn across a blue sky, branching feather-like plumes and curved lines ending in tufts. These lines are often arranged in bands which cross the sky in lines and which, owing to the effect of perspective, appear to converge to a point on the horizon, or to two opposite points (i.e. polar bands). Cirrostratus and cirrocumulus often take part in the formation of these bands. Before sunrise and after sunset, cirrus is sometimes coloured bright yellow or red. Owing to their great height cirriform clouds are illuminated long before other clouds and fade out much later. Observation of cirrus at night is difficult but, if thick and extensive, it may be noted by its dimming effect on stars.

---

*See footnote on page 51.
†See after page 56 for cloud photographs. It will be noted that these are arranged in order of 'Cloud type' according to the specifications of the code for reporting cloud. (Pages 50–52.)

46

Cirrocumulus (Cc). A cirriform layer or patch composed of small white flakes or of very small globular masses, without shadows, which are arranged in groups or lines, or more often in ripples resembling those of the sand on the sea-shore.

In general, cirrocumulus represents a degraded state of cirrus and cirrostratus, both of which may change into it. In this case the changing patches often retain some fibrous structures in places. Real cirrocumulus is uncommon. It must not be confused with small altocumulus on the edges of altocumulus sheets. In the absence of any other criterion the term cirrocumulus should only be used when:

(a) there is evident connection with cirrus or cirrostratus,

or (b) the cloud observed results from a change in cirrus or cirrostratus.

Cirrostratus (Cs). A thin whitish veil, which does not blur the outlines of the sun or moon, but gives rise to haloes. Sometimes it is quite diffuse and merely gives the sky a milky look; sometimes it more or less distinctly shows a fibrous structure with disordered filaments. Cirrostratus may be observed at night by noting the slight diffusion of light around each star, whose brilliance is at the same time dimmed. It is almost impossible to differentiate between thick cirrus and cirrostratus at night in the absence of moonlight.

Altocumulus (Ac). A layer or patches, composed of laminae or rather flattened globular masses, the smallest elements of the regularly arranged layers being fairly small and thin, with or without shading. These elements are arranged in groups, in lines, or waves, following one or two directions and are sometimes so close together that their edges join.

When the edge or a thin translucent patch of altocumulus passes in front of the sun or moon a corona appears. This phenomenon may also occur with cirrocumulus and with the higher forms of stratocumulus. Irisation or iridescence is another possibility with altocumulus. (See also pages 97 and 106.)

The limits within which altocumulus is met are very wide. At the greatest heights, when made up of small elements, it resembles cirrocumulus; altocumulus, however, is distinguished by not being either closely associated with cirrus or cirrostratus or evolved from one of these types. It is often associated with altostratus and either form may change into the other.

Two important varieties of altocumulus are 'altocumulus castellanus' and 'altocumulus lenticularis'. Altocumulus castellanus is a variety peculiar to a thundery state of the atmosphere, and is sure evidence of high-level instability. In this form individual cloudlets are extended vertically upwards in heads or towers, like small cumuli. The lenticular variety of altocumulus has clouds of an ovoid or lens shape, with clear-cut edges and sometimes showing irisations. It occurs frequently over mountainous country and in 'föhn', 'scirocco' and 'mistral' winds. It may also often be seen after the passage of weak cold fronts.

Altostratus (As). Striated or fibrous veil, more or less grey or bluish in colour. This cloud is like thick cirrostratus, but does not show halo phenomena; the sun or moon shows vaguely, with a gleam, as though through ground glass. Sometimes the sheet is thin with forms intermediate with cirrostratus. Sometimes it is very thick and dark, perhaps even completely obscuring the sun or moon. In this case differences of thickness may cause relatively light patches between very dark parts; but the surface never shows real relief, and the striated or fibrous structure is always seen in places in the body of the cloud. Every gradation is observed between high altostratus and cirrostratus on the one hand and low altostratus and nimbostratus on the other. In practice it is

47

important to distinguish between altostratus (thin) through which the sun or moon is visible and altostratus (thick) which completely obscures the sun or moon.

Nimbostratus (Ns). A low, amorphous (i.e. without form), and rainy layer, of a dark grey colour and nearly uniform; feebly illuminated seemingly from inside. Precipitation from nimbostratus is nearly always 'continuous'; but precipitation is not a sufficient criterion. Cloud may be described as nimbostratus before precipitation has started. There is often precipitation which does not reach the ground; in this case the base of the cloud is always diffuse and looks 'wet' on account of the general trailing precipitation, 'virga',* so that it is not possible to determine precisely the limit of its lower surface.

Nimbostratus is usually the result of a progressive lowering and thickening of a layer of altostratus. Beneath nimbostratus there is generally a progressive development of very low ragged clouds (scud). These clouds are usually referred to as **stratus fractus (St fra).**

Stratus (St). A uniform layer of cloud, resembling fog but not resting on the ground. When this very low layer is broken up into irregular shreds it is designated stratus fractus (St fra). A veil of true stratus generally gives the sky a hazy appearance which is very characteristic, but which in certain cases may cause confusion with nimbostratus. When there is precipitation the difference is manifest; stratus cannot give the continuous precipitation usually associated with nimbostratus. When there is no precipitation a dark and uniform layer of stratus can easily be mistaken for nimbostratus. The lower surface of nimbostratus, however, always has a wet appearance (widespread trailing precipitation or virga); it is quite uniform and it is not possible to make out definite details. Stratus on the other hand has a 'drier' appearance and, however uniform it may be, it shows some contrasts and some lighter transparent parts. Stratus is often a local cloud and, when it breaks up, the blue sky is often seen.

A common mode of stratus formation is the slow lifting of a fog layer due to increase in wind speed or warming of the surface.

Stratocumulus (Sc). A layer or patches composed of rounded masses or rolls; the smallest of the regularly arranged elements are fairly large; they are soft and grey, with darker parts. These elements are arranged in groups, in lines, or in waves, aligned in one or two directions. Very often the rolls are so close that their edges join together; when they cover the whole sky as on the continent, especially in winter, they have a wavy appearance. The difference between stratocumulus and altocumulus is essentially one of height. A cloud sheet called altocumulus by an observer at a lower height may appear as stratocumulus to an observer at a considerably greater height.

Stratocumulus may form by the spreading out of cumulus. This happens over land in the evening when the day-time cumulus clouds begin to spread out prior to dissolving. Another example is when developing cumulus meets a pronounced inversion layer. If unable to penetrate this layer the cloud spreads out horizontally in the form of stratocumulus.

Cumulus (Cu). Thick clouds with vertical development; the upper surface is dome-shaped and exhibits rounded protuberances, while the base is nearly horizontal. When the cloud is opposite to the sun the surfaces normal to the

---

*Latin; Virga—streak, bough or broom.

observer are brighter than the edges of the protuberances. When the light comes from the side the clouds exhibit strong contrasts of light and shade; against the sun, on the other hand, they look dark with a bright edge. True cumulus is definitely limited above and below, and its surface often appears hard and clear-cut; but one may also observe a cloud resembling ragged cumulus in which the different parts show constant change. This cloud is called **cumulus fractus (Cu fra).** Cumulus, whose base is horizontal, clear-cut and generally of a grey colour, has a uniform structure, that is to say it is composed of rounded parts right up to its summit, with no fibrous structure. One of the species of cumulus, cumulus congestus, can produce abundant precipitation in the tropics. As cumulonimbus generally results from development and transformation of cumulus, it is sometimes difficult to distinguish cumulus with great vertical extent from cumulonimbus. If it is not possible to decide on the basis of other criteria, the cloud should, by convention, be called cumulus if it is not accompanied by thunder, lightning or hail.

Cumulus having but small vertical development and little individual extent is known as 'fair-weather cumulus' to distinguish it from the ordinary 'large cumulus'.

Cumulonimbus (Cb). Heavy masses of cloud, with great vertical development, whose cumuliform summits rise in the form of mountains or towers, the upper parts having a fibrous texture and often spreading out in the shape of an anvil. The base of the cloud resembles nimbostratus, and one generally notices 'virga' (trailing precipitation). This base has often a layer of very low ragged clouds below it.

Cumulonimbus clouds generally produce showers of rain or snow, and sometimes of hail or soft hail, and often thunderstorms as well. If the whole of the cloud cannot be seen, the fall of a moderate or heavy shower is enough to characterize the cloud as a cumulonimbus. A cumulonimbus cloud may cover the whole sky, in which case the base alone is visible and resembles nimbostratus from which it is difficult to distinguish it. If the cloud mass does not cover all the sky and if even small portions of the upper parts of the cumulonimbus appear, the difference is evident. In other cases the distinction can only be made if the preceding evolution of the clouds has been followed or if precipitation occurs. Cumulonimbus gives showers whereas nimbostratus is associated with continuous precipitation. When there is doubt as to the choice between cumulonimbus and cumulus, cumulonimbus should be reported if it is accompanied by lightning, thunder, or hail.

The lower surface of cumulonimbus sometimes has an udder-like or mamillated appearance which is referred to as 'mamma'. When a layer of menacing cloud covers the sky and mammatus structure and trailing precipitation are both seen it is a sure sign that the cloud is the base of a cumulonimbus, even in the absence of all other signs.

The clouds which develop in the rear of depressions are often cumulonimbus. However, by the spreading out of the upper parts of this cloud and the dissolving of the lower parts, altocumulus or stratocumulus can form. Dense cirrus will develop when the cirriform upper parts of cumulonimbus spread out.

Making the observations. The aspect of the sky is continually changing and the cloud formations in evidence at one particular time may not be typical, that is to say they may not be easily recognizable from the standard descriptions given above. If, however, the observer watches the sky over a period of time he will often find that doubtful cloud forms may be referred to a previous state of development that was typical. Hence the first rule in cloud observing—watch the sky as often as possible and not merely at the time of observation.

Coding the observations. The forecaster who eventually receives and uses the observer's reports does not merely want to know what clouds are present. It has been found that certain distributions or arrangements of clouds in the sky, in other words certain 'states of sky', are of particular significance. The observer is required to report these rather than the presence of a particular cloud form. These states of sky are as follows, separate specifications being used for low, medium and high cloud.

SPECIFICATION OF FORM OF LOW CLOUD (C_L) (Sc, St, Cu, Cb)

*Code
figure*

0   No Stratocumulus, Stratus, Cumulus or Cumulonimbus.

1   Cumulus with little vertical extent and seemingly flattened, or ragged Cumulus other than of bad weather,* or both.

2   Cumulus of moderate or strong vertical extent, generally with protuberances in the form of domes or towers, either accompanied or not by other Cumulus or by Stratocumulus, all having their bases at the same level.

3   Cumulonimbus the summits of which, at least partially, lack sharp outlines, but are neither clearly fibrous (cirriform) nor in the form of an anvil; Cumulus, Stratocumulus or Stratus may also be present.

4   Stratocumulus formed by the spreading out of Cumulus; Cumulus may also be present.

5   Stratocumulus not resulting from the spreading out of Cumulus.

6   Stratus in a more or less continuous sheet or layer, or in ragged shreds, or both, but no Stratus fractus of bad weather.*

7   Stratus fractus of bad weather* or Cumulus fractus of bad weather, or both, usually below Altostratus or Nimbostratus.

8   Cumulus and Stratocumulus other than that formed from the spreading out of Cumulus; the base of the Cumulus is at a different level from that of the Stratocumulus.

9   Cumulonimbus, the upper part of which is clearly fibrous (cirriform), often in the form of an anvil; either accompanied or not by Cumulonimbus without anvil or fibrous upper part, by Cumulus, Stratocumulus, Stratus or Stratus fractus.

/   Stratocumulus, Stratus, Cumulus and Cumulonimbus invisible owing to darkness, fog, blowing dust or sand, or other similar phenomena.

SPECIFICATION OF FORM OF MEDIUM CLOUD (C_M) (Ac, As, Ns)

*Code
figure*

0   No Altocumulus, Altostratus or Nimbostratus.

1   Altostratus, the greater part of which is semi-transparent; through this part the sun or moon may be weakly visible, as through ground glass.

---

*'Bad weather' denotes the conditions which generally exist during precipitation and a short time before and after.

2   Altostratus, the greater part of which is sufficiently dense to hide the sun or moon, or Nimbostratus.*

3   Altocumulus, the greater part of which is semi-transparent; the various elements of the cloud change only slowly and are all at a single level.

4   Patches (often in the form of almonds or fishes) of Altocumulus, the greater part of which is semi-transparent; the clouds occur at one or more levels and the elements are continually changing in appearance.

5   Semi-transparent Altocumulus in bands, or Altocumulus in one or more fairly continuous layers (semi-transparent or opaque), progressively invading the sky; these Altocumulus clouds generally thicken as a whole.

6   Altocumulus resulting from the spreading out of Cumulus (or Cumulonimbus).

7   Altocumulus in two or more layers, usually opaque in places, and not progressively invading the sky; or opaque layer of Altocumulus, not progressively invading the sky; or Altocumulus together with Altostratus or Nimbostratus.*

8   Altocumulus with sproutings in the form of small towers or battlements, or Altocumulus having the appearance of cumuliform tufts.

9   Altocumulus of a chaotic sky, generally at several levels.

/   Altocumulus, Altostratus and Nimbostratus invisible owing to darkness, fog, blowing dust or sand or other similar phenomena, or more often because of the presence of a continuous layer of lower clouds.

SPECIFICATION OF FORM OF HIGH CLOUD (C_H) (Ci, Cs, Cc)

Code figure

0   No Cirrus, Cirrocumulus or Cirrostratus.

1   Cirrus in the form of filaments, strands or hooks, not progressively invading the sky.

2   Dense Cirrus, in patches or entangled sheaves, which usually do not increase and sometimes seem to be the remains of the upper part of a Cumulonimbus; or Cirrus with sproutings in the form of small turrets or battlements, or Cirrus having the appearance of cumuliform tufts.

3   Dense Cirrus, often in the form of an anvil, being the remains of the upper parts of Cumulonimbus.

4   Cirrus in the form of hooks or of filaments, or both, progressively invading the sky; they generally become denser as a whole.

5   Cirrus (often in bands converging towards one point or two opposite points of the horizon) and Cirrostratus, or Cirrostratus alone; in either case, they are progressively invading the sky, and generally growing denser as a whole, but the continuous veil does not reach 45 degrees above the horizon.

---

*For synoptic purposes nimbostratus is included among the medium clouds in the code since it is continuous with the altostratus existing above it and has been formed as a result of a progressive lowering of altostratus from medium cloud level.

6   Cirrus (often in bands converging towards one point or two opposite points of the horizon) and Cirrostratus, or Cirrostratus alone; in either case, they are progressively invading the sky, and generally growing denser as a whole; the continuous veil extends more than 45 degrees above the horizon, without the sky being totally covered.

7   Veil of Cirrostratus covering the celestial dome.

8   Cirrostratus not progressively invading the sky and not completely covering the celestial dome.

9   Cirrocumulus alone, or Cirrocumulus accompanied by Cirrus or Cirrostratus, or both, but Cirrocumulus is predominant.

/   Cirrus, Cirrocumulus and Cirrostratus invisible owing to darkness, fog, blowing dust or sand or other similar phenomena, or more often because of the presence of a continuous layer of lower clouds.

The use of these specifications, instead of reporting each individual cloud form, effects an economy and is also advantageous to the forecaster, who knows how to associate a state of sky with a particular weather situation.

The estimation of cloud height. Apart from some ships of the Royal Navy and occasionally on ocean weather ships, cloud height at sea is obtained by estimation. The first step in estimating cloud height consists of identifying the cloud as a type belonging to one of the three classes, low, medium or high. Low clouds have their bases below 6500 feet (2000 m). Medium cloud layers usually occur at levels between 6500 and 18 000 feet (2000 and 5500 m), and high clouds are usually above 18 000 feet (5500 m). As a rough guide, the heights of the bases of the various types of low cloud may be expected to be between the following limits:

Stratus	Usually below 2000 feet (600 m) and sometimes nearly down to the surface.
Nimbostratus	500 to 4000 feet (150 to 1200 m), usually below 2000 feet (600 m) in moderate rain or snow.
Cumulonimbus	2000 to 5000 feet (600 to 1500 m).
Stratocumulus	1500 to 4500 feet (450 to 1350 m).
Cumulus	1500 to 5000 feet (450 to 1500 m).

These limits tend to be considerably higher in low latitudes; this applies particularly to high clouds.

It is difficult to estimate cloud height without much practice. The apparent size of the cloud elements is often an indication of height. For example, the lower the height of the individual cloudlets of an altocumulus layer, the larger they will normally appear. Layers having the appearance of altocumulus with large individual elements are often found at heights between 6000 feet (1800 m) and 10 000 feet (3000 m). The estimation of the height of stratified cloud, e.g. altostratus or nimbostratus, is particularly difficult. The lack of pronounced structure makes it easy to gain a false impression of height. Valuable experience can be gained on occasions when the observer knows that his ship is steaming towards a depression by watching the gradual lowering of the cloud base. The observer's impressions of the appearance of the sky in the successive stages of lowering will assist his judgment on future occasions. It is only by such experience that an observer can distinguish between a layer of nimbostratus in the lower middle band and a similar layer at, perhaps, only 2000 to 3000 feet (600 to 900 m).

Care must be taken before using the apparent speed of cloud as an index to its height. This apparent speed depends not only on the velocity of the wind at cloud level but also on the course and speed of the ship itself.

When coasting, cloud height may sometimes be estimated by comparison with the height of the mountains or hills in the background. In using this method, however, it should be remembered that cloud is usually lower over the hills than elsewhere and that it is the general level over the sea that is required.

The estimation of cloud amount. The amount of cloud was in the past estimated as the number of tenths of sky covered. At a conference of the International Meteorological Organization (Washington 1947) it was recommended that amount of cloud be reported in eighths instead of tenths. This change of procedure was brought into force with the introduction of the revised International Code (Washington) on 1 January 1949.

In making the observation it is necesary to stand in a position affording an uninterrupted view of the whole sky. To make an estimate for the whole sky at once requires practice and is rather difficult at first. It is convenient to imagine the sky divided into quadrants by two arcs drawn at right angles through the zenith.

Each quadrant represents two-eighths of the total sky. If we choose the most appropriate of the figures—

0 = Clear or almost clear of cloud
1 = About half covered
2 = Completely or almost completely covered with cloud—

for each separate quadrant, then the total amount of cloud for the whole sky is obtained simply by adding the amounts in the separate quadrants.

At night the observation of total cloud amount is noted by observing which stars are showing and which are obscured. It is more difficult to differentiate between low, middle and high clouds and reliable observation depends upon the degree of illumination and the experience of the observer.

53

CHAPTER 6

Ocean Waves

The complex nature of wave motion at sea. The action of wind in producing waves is not precisely understood. The effect of the wind varies from the tiny ripples ruffled on a pond by the merest breath of air to the mighty rollers of the North Atlantic and Roaring Forties. All ocean waves, other than those caused by movements of the sea floor and tidal effects, owe their origin to the generating action of the wind. Wave motion, however, may persist even after the generating force has disappeared, being then slowly dissipated by frictional forces.

An observer of the motion of the sea surface at a particular place will, in general, notice a complicated wave form such as is shown in Figure 27 (page 58), which may be regarded as the result of the superposition of a number of simple regular wave motions having different lengths and speeds.

The ideal observer is an instrument known as a wave-recorder which registers automatically the up and down motion of the water surface and enables a record such as Figure 27 to be drawn. This record can be analysed or split up into its component simple waves. Most wave-recorders can only be effectively used from the shore, offshore installations or from stationary ships and hence it is not possible to measure sea disturbance in general by this method although it would be most desirable to do so.

The distinction between sea and swell. The system of waves raised by the local wind blowing at the time of observation is usually referred to as 'sea'. Those waves not raised by the local wind blowing at the time of observation, but due either to winds blowing at a distance or to winds that have ceased to blow, are known collectively as 'swell'. Usually, one component of the swell dominates the rest, but occasionally two component wave motions crossing at an angle may be observed. These are referred to as 'cross swells'. Sea and swell may both be present at the same time and the sea may be from a different direction and have different period and height to the swell, or both sea and swell may be from the same direction.

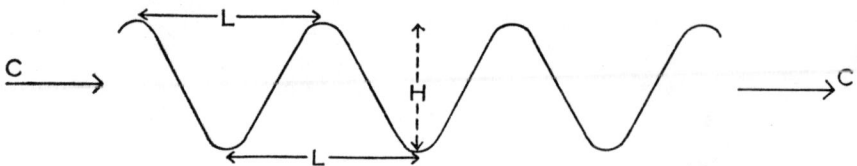

FIGURE 24. Characteristics of a simple wave

The characteristics of a simple wave. The following definitions are used in describing a simple wave:

(a) SPEED, C, usually expressed in knots, is the speed at which individual waves travel.

(b) LENGTH, L, expressed in metres, is the horizontal distance between successive crests or successive troughs.

54

(c) PERIOD, T, expressed in seconds, is the time interval required for the passage of successive crests (or successive troughs) past a given point.

(d) HEIGHT, H, expressed in metres, is the vertical distance between the top of a crest and the bottom of a trough.

The following relations are found to hold for a simple wave:

Speed $\; = 3 \cdot 1 \times$ period
Length $= 1 \cdot 555 \times$ (period)^2 metres.

(In application to actual sea waves, which are not simple, the constant $1 \cdot 555$ should be reduced by a factor ranging between about $\frac{1}{2}$ and $\frac{2}{3}$.)

By means of these formulae, measurements of one of the variables can be used to calculate the other two. The following table gives these relations numerically for different wave periods:

Period seconds	Length metres	Speed knots
2	6·2	6·2
4	24·9	12·4
6	56·0	18·6
8	99·5	24·8
10	155·5	31·0
12	223·9	37·2
14	304·8	43·4
16	398·0	49·6
18	503·9	55·8
20	622·0	62·0

There is no inherent theoretical relation between the height and period of a simple wave. We can imagine the height to be varied at will, the period (and hence length and speed) remaining constant. In real wave motion, however, in which many simple waves are superposed there is a further consideration that enables us to see how the height is limited. If we call the quotient H/L the 'steepness' of the wave, it is found that the mean steepness does not increase beyond 7·6 per cent (1/13). If the mean steepness is less than this figure then the waves are capable of absorbing more energy from the wind, thus increasing their height relative to their length. When the limiting steepness is reached, surplus energy received from the wind is dissipated by the breaking of the waves at the crests (white horses). This limiting value of the steepness explains why the mean maximum height of the sea waves is roughly in proportion to their length; for example, wind-driven waves of length 120 m (period 9 seconds) would not be expected to have a mean maximum height greater than 9 m. If the wave-length were about 150 m (period 10 seconds) this limiting value of the mean maximum height would be increased to 12 m. On the other hand, long swells, perhaps 300–600 m in length, may have heights of less than half a metre.

When the height of the wave is small compared with its length, the wave profile can be adequately represented by a simple sine curve. As the height becomes relatively greater, however, it is seen that the crests become sharper and the troughs much more rounded, the precise profile being a curve known as a 'trochoid'. This is the curve that would be traced on a bulkhead by a marking point fixed to the spoke of a wheel, if we imagined the wheel to be rolled along under the deckhead.

55

In Figure 25 the large circle represents the wheel and P the marking point on a spoke, OP, the distance from the axle being called the tracing arm. The arrow shows the direction in which the circle rolls and in which the wave is supposed to be travelling. AB is the base, i.e. the straight line under which the circle is to roll, the length AB being equal to the half circumference of the wheel, AR.

Now as the circle rolls, when position 3 of the circle reaches position 3 of the base, the semi-circle FPG will be in the position shown by the dotted semi-circle; and the marking point P will coincide with the point D, having described part of a trochoid PD. When the circle has completed half a revolution, the marking point P will coincide with E, having described the trochoid curve PDE which is half a wave-length; the diameter POH represents the height of the wave. The nearer the marking point is to the axle of the wheel, the flatter will be the trochoid.

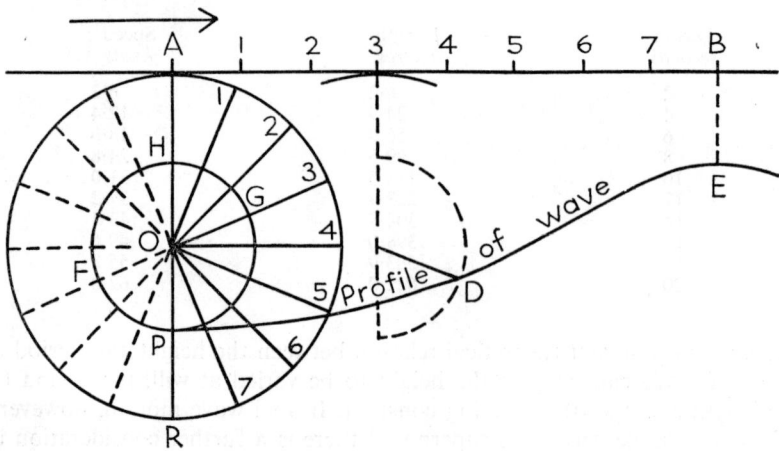

FIGURE 25. Representation of a trochoidal wave form

In an ideal wave each water particle revolves with uniform speed in a circular orbit, perpendicular to the wave ridge (the diameter of the orbital circles being the height of the wave) and completes a revolution in the same time as the wave takes to advance its own length. At a wave-crest the motion of the particles is wholly horizontal, advancing in the same direction as the wave; at mid height on the front slope it is wholly upwards; in the trough it is again horizontal but in the opposite direction to the travel of the wave, and at mid height on the back slope it is wholly downwards. This motion may be seen by watching a floating object at the passage of a wave. The object describes a circle but is not carried bodily forward by the wave.

The disturbance set up by wave motion must necessarily extend for some distance below the surface; but its magnitude decreases very rapidly in accordance with a definite law, the trochoids becoming flatter and flatter as the depth increases, and the water particles revolving in ever-decreasing circles. At a depth of one wave-length the disturbance is less than a five-hundredth part of what is is at the surface, so that the water at that depth may be considered undisturbed. The motion associated with the largest ocean waves is inappreciable at even moderate depths, as is demonstrated by experience in submarines.

Eilean Ban, Kyle of Lochalsh   *Photograph by R. K. Pilsbury*

Photograph by C. E. Wallington

C$_L$1 Cumulus with little vertical extent. The cloud elements shown in the upper picture are in an early stage of development; they are small, shallow and have ragged edges. In the lower picture they are in a slightly more advanced stage and some show the characteristic domed tops

CLOUD PLATE I

G

From Dale, Milford Haven   *Photograph by R. K. Pilsbury*

C_L2 **Cumulus of moderate or strong vertical extent.** This is a further stage in the development of C_L1. The cloud has become much deeper and the tops are 'cauliflower shaped'. The outlines are clear cut and there is no tendency for the upper parts of the cloud mass to become blurred or fibrous in texture. When the cloud is well developed, rain showers may occur. Stratocumulus cloud may also be present but it must be at the same level as the base of the Cumulus.

Near Tenby, Dyfed   *Photograph by R. K. Pilsbury*

C_L3 **Cumulonimbus without anvil.** Normally this is a further stage in the development of C_L2. The tops are beginning to acquire a fibrous appearance. When it is uncertain whether the cloud is C_L3 or C_L9, the latter should be selected if the cloud gives rise to lightning, thunder or hail. Cumulus, Stratocumulus or Stratus may also be present.

CLOUD PLATE II

Photograph by R. K. Pilsbury

C_L4 Stratocumulus formed by the spreading out of Cumulus. Cloud of this type forms when the upper parts of Cumulus clouds, which had previously been gaining height, can no longer do so and begin to spread out horizontally, forming a layer of Stratocumulus, as illustrated above. (See Plate XIV describing the formation of C_M6.) Sometimes the spreading out is only temporary and the Cumulus resumes growth above the stable layer.

Another type of C_L4 often occurs in the evening when convection ceases and, in consequence, the Cumulus begins to flatten, and assumes the appearance of patches of Stratocumulus.

CLOUD PLATE III

From Malahide, near Dublin   *Photograph by R. K. Pilsbury*

Lough Gill, Co. Sligo   *Photograph by R. K. Pilsbury*

C$_L$5 Stratocumulus not formed by the spreading out of Cumulus. The individual cloud masses may be separate and in the form of elongated bands as shown in the upper picture, or they may be closed up into a continuous or nearly continuous layer, as shown in the lower photograph. Often the cloud is dark and heavy looking, but it can be light in tone when it is at a fairly high level, or when it is thin.

CLOUD PLATE IV

Benbulbin, Co. Sligo   *Photograph by R. K. Pilsbury*

North Antrim   *Photograph by R. K. Pilsbury*

C_L6 **Stratus in a more or less continuous sheet or layer,** or in ragged shreds, or both. The upper picture shows a layer of Stratus with its base about 350 feet above ground level, 1100 feet above sea level. In the lower picture the ragged Stratus was formed from lifted sea fog.

CLOUD PLATE V

Photograph by G. A. Clarke

Photograph by R. K. Pilsbury

C$_L$7 Stratus fractus of bad weather, or ragged Cumulus, or both, generally moving fast and changing shape rapidly. These clouds (known to mariners as 'scud') often form beneath a layer of Nimbostratus, as in the upper picture, or a layer of Altostratus, as in the lower picture.

CLOUD PLATE VI

Bridlington Bay   *Photograph by R. K. Pilsbury*

Photograph by R. K. Pilsbury

$C_L 8$ **Cumulus and Stratocumulus,** other than that formed from the spreading out of Cumulus, with their bases at different levels. The base of the Cumulus is normally at the lower level and occasionally the tops may reach or penetrate the Stratocumulus as illustrated above.

CLOUD PLATE VII

Photograph by D. E. Watts

Photograph by K. J. Norris

C_L9 **Cumulonimbus with anvil.** This is a massive cloud of great vertical depth and horizontal extent, having a frayed-out fibrous top in the shape of an anvil. It normally develops from C_L3 and commonly gives rise to thunderstorms and/or hail showers. Underneath the base of the cloud, which is often very dark, there are frequent low, ragged clouds, which in storms are only a few hundred feet above the surface. Occasionally the upper parts of the cloud merge with Altostratus or Nimbostratus. The lower picture shows the appearance which the base of a Cumulonimbus cloud presents when it is overhead. The downward-hanging protuberances are due to turbulent downdraughts and are known as 'mamma'.

CLOUD PLATE VIII

Photograph by R. K. Pilsbury

C_M1 **Thin Altostratus.** The background cloud seen in the picture has probably developed from the thickening Cirrostratus associated with an approaching warm front. Most of the cloud sheet is light grey and semi-transparent with the sun shining weakly through it. The cloud below the Altostratus is Stratocumulus (C_L5).

CLOUD PLATE IX

Caldy Island, Carmarthen Bay   *Photograph by R. K. Pilsbury*

Photograph by R. K. Pilsbury

C$_M$2 Thick Altostratus or Nimbostratus. This develops from thickening C$_M$1. If only the denser parts of the cloud hide the sun or moon it is defined as Altostratus; if the cloud is sufficiently opaque throughout to obscure them completely, it is Nimbostratus, as shown in the lower picture (with Stratus fractus beneath the cloud layer).

CLOUD PLATE X

Photograph by R. K. Pilsbury

Photograph by C. J. P. Cave

C_M3 **Semi-transparent Altocumulus,** at a single level, not invading the sky. The elements in the cloud sheet are small, rounded and more or less uniform; near the sun they are translucent. The sky may contain several Altocumulus patches of different opacity, as in the upper picture, or thin, lightly shaded sections as in the lower picture.

CLOUD PLATE XI

Photograph by G. A. Clarke

C_M4 **Altocumulus lenticularis.** The clouds shown in the picture are due to wave motion in the atmosphere and they are seen mainly over hilly country. At sea they are therefore likely to be seen only in certain coastal waters in the direction of the land. The cloud elements are smooth looking and taper away towards the ends; they have bright translucent edges and show definite shading. In another variety of the cloud the elements are composed of fine granules and ripples lying in thin irregular patches, vaguely lenticular in shape and having fairly pronounced shading. Parts of the cloud near the sun often show the delicate colouring known as irisation.

CLOUD PLATE XII

Photograph by R. K. Pilsbury

C_M5 **Altocumulus increasing and thickening.** The essential feature of this cloud type is that it spreads from some particular direction on the horizon and increases in amount, perhaps finally covering the whole sky. It may be in one or more layers of varying degrees of density, some parts being translucent and others heavily shaded. Sometimes it may resemble Stratocumulus at a high level.

CLOUD PLATE XIII

C$_M$6 Altocumulus formed by the spreading out of Cumulus or Cumulonimbus. In certain atmospheric conditions,* the tops of these clouds reach a level above which they cannot rise, and they are compelled to flatten out horizontally, forming patches of Altocumulus of rather irregular thickness and shape, as shown in the picture. (Care should be taken not to confuse these with the anvil-shaped tops of Cumulonimbus.) If many large Cumulus-type clouds are present, the patches of Altocumulus, resulting from the spreading out of their tops, may coalesce to form quite an extensive layer.

* *When there is a marked temperature inversion in the atmosphere it acts as a barrier to air rising by convection.*

$C_M 7$ **Altocumulus not increasing.** This type includes:
    (a) Patches, sheets or layers of Altocumulus at different levels, not increasing.
    (b) Patches, sheets or layers of Altocumulus at a single level, generally opaque, not
        increasing.
    (c) Altocumulus, together with Altostratus or Nimbostratus.
    The upper picture shows type (c) with the Altocumulus merging with Altostratus near the
horizon. Beneath the thick patch of Altocumulus, type (b), in the lower picture can be seen small
pendant globules called 'mamma'.

CLOUD PLATE XV

Photograph by R. K. Pilsbury

Photograph by C. J. P. Cave

C_M8 **Altocumulus castellanus and Altocumulus floccus.** Both these types are associated with developing thundery conditions over a wide area as opposed to thunderstorms arising from locally generated Cumulonimbus clouds. The upper picture shows a long line of typical Altocumulus castellanus in the distance; in the top half of the picture there are lines of Altocumulus in cumuliform tufts (Altocumulus floccus). The lower picture shows very ragged Altocumulus floccus in considerable quantity.

CLOUD PLATE XVI

Skye from Sound of Sleat   *Photograph by R. K. Pilsbury*

C_M9 **Altocumulus of a chaotic sky,** generally at several levels. The main characteristic of the sky is its heavy and stagnant appearance. There are usually clouds of all medium types mixed with clouds of type C_L and C_H.

Photograph by R. K. Pilsbury

C_H1 **Cirrus in hooks or filaments,** not progressively invading the sky. The picture clearly shows the fibrous structure of Cirrus, with hooks at the ends of the fine strands. This type of Cirrus often occurs with other Cirrus clouds and should be classified as C_H1 only when the combined cover of all filaments, strands or hooks exceeds the total of other types of Cirrus present.

CLOUD PLATE XVIII

Photograph by R. K. Pilsbury

Photograph by R. K. Pilsbury

C_H2 **Dense Cirrus.** The Cirrus occurs in dense patches or entangled sheaves (upper picture), not usually increasing in amount. The two dense patches in the centre of the lower picture are typical examples. Sometimes the Cirrus is arranged in narrow bands, with sproutings in the shape of small turrets or battlements, or showing cumuliform tufts.

CLOUD PLATE XIX

Photograph by R. K. Pilsbury

C_H3 **Dense Cirrus from Cumulonimbus,** often in the form of an anvil. These clouds are white and fibrous and may be entirely separate from the Cumulonimbus of which they were a part.

Photograph by R. K. Pilsbury

C_H4 **Cirrus invading the sky.** These clouds can be seen as separate, delicate filaments with fibrous trails. The Cirrus elements in the picture are moving from left to right, invading the sky and thickening, but no Cirrostratus is present.

CLOUD PLATE XX

Photograph by R. K. Pilsbury

C_H5 Cirrus and/or Cirrostratus below 45° altitude. A whitish veil progressively invading the sky but less than 45° above the horizon from which the cloud is spreading. Cirrus of this type develops as a result of the continued spread of C_H4. The clouds in the upper part of the picture are Cirrus. Nearer the horizon the Cirrus has merged into a sheet of Cirrostratus.

CLOUD PLATE XXI

Near Clovelly, looking north   *Photograph by R. K. Pilsbury*

C$_H$6 Cirrus and/or Cirrostratus above 45° altitude. The picture shows Cirrus thickening rapidly to Cirrostratus, already extending higher than 45° above the horizon. The main sheet of Cirrostratus is often preceded by long filaments or hooks of Cirrus, sometimes in bands across the sky.

CLOUD PLATE XXII

Photograph by R. K. Pilsbury

C_H7 **Veil of Cirrostratus covering the whole sky.** The cloud is usually light, uniform and nebulous, but it may be white and fibrous showing striations. It gives rise to halo phenomena round the sun or moon. Sometimes the veil is so thin that it is difficult to distinguish it from the blue sky, and halo phenomena may provide the only reliable evidence of its presence.

Photograph by G. A. Clarke

C_H8 **Cirrostratus not increasing and not covering the whole sky.** The veil or sheet of clouds extends from one horizon but leaves a segment of blue sky in the other direction. The cloud either remains constant in amount or decreases. The picture shows a fairly dense sheet of Cirrostratus clearing away after the passage of a cold front.

CLOUD PLATE XXIII

Photograph by R. K. Pilsbury

Photograph by R. K. Pilsbury

C$_H$9 Cirrocumulus. This code figure is only used when the amount of Cirrocumulus is greater than any other types of Cirrus which may be present. The upper picture shows a typical example in the centre of the picture, the cloud elements being very much smaller than those found in Altocumulus. A few Cumulus clouds are also present. In the lower picture the elements are composed of small, ragged cumuliform tufts.

CLOUD PLATE XXIV

Wave groups. Experience shows that waves generally travel in groups with patches of dead water in between, the wave height being a maximum at the centre of each group. We have said earlier that any observed wave motion can be regarded as built up from a number of simple wave forms. Let us consider, for example, the superposition of two simple wave motions having the same height but slightly different periods. If the crests of the two wave motions are made to coincide at the initial point of observation the height of the resultant wave will be twice that of each component wave. To each side of this point, however, owing to the difference of period, the additive effect becomes less until a point is reached where the heights of the component waves, being of different sign, completely annul each other's effect. Beyond this point the heights again become additive until the troughs of the component waves coincide. In other words, there is a variation of height superposed on the ordinary wave motion. It can also be shown that two simple wave trains moving in slightly different directions give a resultant pattern composed of 'short-crested' waves as distinct from the 'long-crested' waves of simple wave motions.

The speed of a wave group is not the same as that of the individual waves composing it. Each individual wave in its turn emerges from the dead water in the rear of the group, travels through the group and subsides in the dead water ahead of it. The speed of the wave group must therefore be less than the speed of an individual wave. Both theoretical considerations and experience show that the wave group travels at one half the speed of the individual waves.

The origin and travel of swell. Swell waves originate in the heavy seas created in a storm area. Short waves have an insufficient store of energy to enable them to travel long distances against the dissipating action of friction. Hence, in general, it follows that swell waves are long waves in comparison with the wind-driven waves at the place of observation.

In calculating the distance travelled by swell, care must be taken to distinguish between the speed of the individual waves and the speed of the wave groups. If, for example, a ship reports the sudden onset of waves whose speed, calculated from the period, is 30 knots, then another ship in the line of advance of these waves will experience their onset at a time obtained by allowing a speed of $\frac{1}{2} \times 30 = 15$ knots for the disturbance.

As swell travels its height decreases. Investigations by the Institute of Oceanographic Sciences show that if R is the distance from the point of generation in nautical miles then the amplitude at distance R is $(\frac{800}{R})^{\frac{1}{4}}$ of that at the point of generation by the wind. Thus, a swell would lose one-half of its height in travelling a distance of 1200 nautical miles. The long swells are the greatest travellers.

Waves in shallow water. All the previous remarks refer to waves in deep water. When a deep-water wave enters shallow waters it undergoes profound modification. Its speed is reduced, its direction of motion may be changed and, finally, its height increases until, on reaching a certain limiting depth, the wave breaks on the shore. Water may be regarded as shallow when the depth is less than half the length of the wave.

The decrease in speed when a wave approaches the shore accounts for the fact that the wave fronts become, in general, parallel to the shore prior to breaking. Figure 26 shows a wave, approaching the shore at an angle, being refracted until it becomes parallel to the shore.

The same reasoning may be applied to explain how waves are able to bend round headlands and to progress into sheltered bays.

H 57

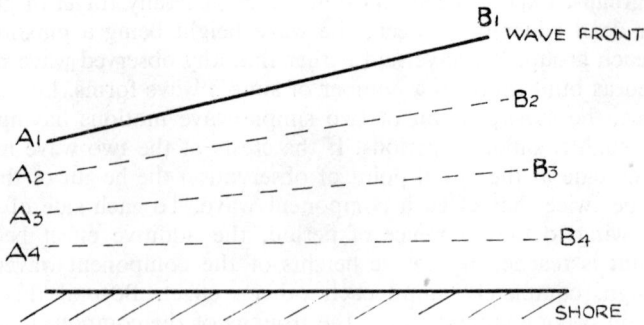

FIGURE 26. Refraction of a wave approaching the shore at an angle

OBSERVING OCEAN WAVES

The inherent difficulties of observation. It has been remarked earlier that the ideal observer is a wave-recorder which can register automatically the up and down motion of the sea surface at a fixed point. A typical record is shown in Figure 27.

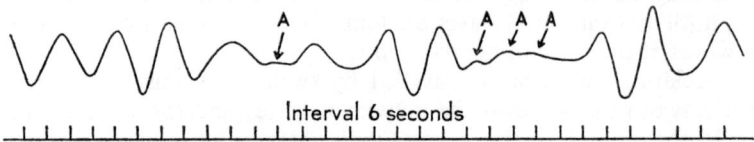

FIGURE 27. Wave form of the sea surface

The record is, in general, complex and shows immediately all the difficulties inherent in eye observation. For example, are all the waves to be considered on an equal footing or are only the big waves to be counted? Since the wave characteristics vary so much, what average values shall be taken? It is obvious that if comparable results are to be obtained the observer must follow a definite procedure. The flat and badly formed waves ('A' in Figure 27) between the wave groups cannot be observed accurately by eye and different observers would undoubtedly get different results if an attempt were made to include them in the record. The method to be adopted, therefore, is to observe only the well-formed waves in the centre of the wave groups. The observation of waves entails the measurement or estimation of the following characteristics:

Direction   Period   Height

Reliable average values of period and height can only be obtained by observing at least twenty waves. Of course, these cannot be consecutive; a few must be selected from each succeeding wave group until the required number has been obtained. Only measurements or quite good estimates are required. Rough guesses have little value and should not be recorded.

58

It will often be found that there are waves coming from more than one direction. For example, there may be a sea caused by the wind then blowing and a swell caused by a wind that has either passed over or is blowing in a distant area. Or there may be two swells (i.e. cross swells) caused by winds blowing from different directions in distant areas. In such cases the observer should distinguish between sea and swell, and report them separately, giving two groups for swell when appropriate.

The direction, height and period of the sea wave may be quite different from that of the swell wave. It will, however, often happen—particularly with winds of Beaufort force 8 and above—that the sea and swell waves are both coming from the same direction. In that case it is virtually impossible to differentiate between sea and swell and the best answer is to look upon the combined wave as being a sea wave and log it accordingly.

Observing waves from a moving ship. Care must be taken to ensure that the observations, especially those of period, are not influenced by the waves generated by the motion of the ship.

(a) DIRECTION FROM WHICH THE WAVES COME. This is easily obtained either by sighting directly across the wave front or by sighting along the crests of the waves and remembering that the required direction differs from this by 90 degrees. Direction is always recorded true, not magnetic.

(b) PERIOD. For measurements of period a stopwatch is desirable. If this is not available an ordinary watch with a seconds hand may be used or, alternatively, a practised observer may count seconds.

The observer selects a distinctive patch of foam or a small object floating on the water at some distance from the ship, and notes the time at which it is on the crest of each successive wave. The procedure is repeated for the larger waves of each successive group until at least twenty observations are available. The period is then taken as the average time for a complete oscillation from crest to crest. In a fast ship it will be found that the 'patch of foam' method will rarely last for more than one complete oscillation and that many waves have to be observed separately. With practice, suitable waves can easily be picked out and the timing from crest to crest becomes quite simple. When it is desired to use an object (an empty beer can is usually conspicuous against the sea and will remain afloat long enough to serve its purpose) it should be thrown as far forward as possible.

Another method available to the observer with a stopwatch is to observe two or more consecutive 'central' waves of a wave group while the watch is running continuously, then to stop the watch until the central waves of the next wave group appear, the watch being then restarted. This procedure is repeated until at least twenty complete oscillations have been observed. The period is then obtained by dividing the total time by the number of oscillations. It is important to note that the periods between times of crests passing a point on the ship are not the ones required.

(c) HEIGHT. Although wave-recorders are fitted to a few research ships, there is at present no method of measuring the height of waves suitable for general use on merchant ships, but a practised observer can make useful estimates. The procedure to be adopted depends on the length of the waves relative to the length of the ship. If the length of the waves is

59

short in comparison with the ship's length, i.e. if the ship spans two or more wave crests, the height should be estimated from the appearance of the waves at or on the side of the ship, at times when the pitching and rolling of the ship is least. For the best result the observer should take up a position as low down the ship as possible, preferably amidships where the effect of pitching is least, and on the side of the ship towards which the waves are coming.

This method fails when the length of the waves exceeds the length of the ship, for then the ship rises bodily with the passage of each wave crest. The observer should then take up a position in the ship so that his eye is just in line with the advancing wave crest and the horizon, when the ship is vertical in the trough. The height of eye above the ship's water line is then the height of the wave. The nearer the observer is to an amidships position the less chance will there be of the measurement being vitiated by pitching. If the ship rolls heavily it is particularly important to make the observation at the moment when she is upright in the trough. Exaggeration of estimates of wave height is mostly due to errors caused by rolling. (See Figure 28. When the ship is rolling (*b*), the observer at 'o' has to take up a higher position to get a line on the horizon than when she is upright (*a*).)

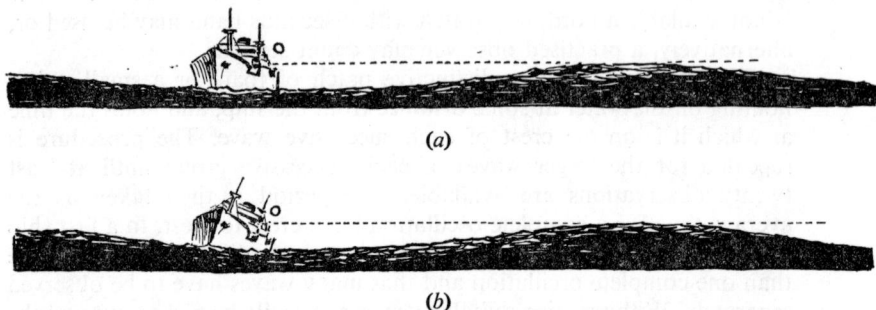

(*a*)

(*b*)

FIGURE 28. Estimation of wave height at sea

The observation of height of waves is most difficult when the length of the waves exceeds the length of the ship and their height is small. The best estimate of height can be obtained by going as near the water as possible, but even then the observation can only be rough. In making height estimates an attempt should be made to fix a standard of height in terms of the height of a man or the height of a bulwark, forecastle or well-known dimension in the ship. There is generally a tendency to overestimate the height of short waves and underestimate the height of long waves.

Estimating the height of a wave from a high bridge in a fast ship is a difficult job and much will depend on the skill and ingenuity of the observer; in many cases all one can hope for is a very rough estimate.

All estimates of wave height should be made preferably with the ship on an even keel so that the observer's height of eye is consistent.

The inherent difficulties already mentioned, together with the practical difficulties of estimation, make it essential that the recorded height be the average value of about twenty distinct observations. These observations should be made on the central waves of the more prominent wave groups.

Wave observations at night or in low visibility. Under these conditions the most that the observer can normally hope to record is direction and an estimate of height, or perhaps direction only, which would at least indicate the presence of waves. Such observations might be of considerable value in tropical waters in the hurricane season. It is only on very bright nights that the observation of period would be practicable.

Observing waves from weather ships. Wave-recorders, which can record the period and height of the waves, have been installed in the British ocean weather ships. But even when no special instruments are carried, weather ships have the advantage of being able to manoeuvre so as to secure the best conditions for wave observation. The methods outlined in (b) may be used to better advantage than by ordinary merchant ships. For example, a floating object may be observed for a considerable time; it is not lost in the distance as occurs when the ship is moving.

In addition to these observations the height, length and period of waves can be determined from a stationary ship as follows:

(a) The estimation of wave height may be much assisted with the use of a dan buoy of known height.
(b) Length can be observed by streaming a buoy for such a distance astern that the crests of successive waves are simultaneously passing the buoy and the observer. The distance between the two is the wave-length.
(c) Period can be obtained by noting the time taken for the wave to travel the distance between the buoy and the observer.

By simple division the speed of the individual waves can be deduced.

The importance of wave observations. The study of ocean waves has only recently been put on a scientific basis by the utilization of an automatic method of recording and the subsequent analysis of the record into component simple waves. The establishment of a network of specially equipped observing stations would probably add much to our present knowledge of the generation, transmission and decay of ocean waves. The new method of recording has made evident the limitations of former methods of observation, including the use of sea and swell scales, and has indicated the necessity of obtaining quantitative observations of wave characteristics.

Of practical importance is the fact that quantitative wave observations may be used for identifying the approximate position of a storm centre when suitable weather observations are lacking. The use of swell as an indication of the approach of a tropical storm is well known. The forecasting of swell on exposed coasts, such as those of Morocco and Portugal, is of considerable value for the protection of coastal shipping and port installations. The accuracy of these forecasts depends largely on an adequate supply of reliable ships' observations. Statistics of the period and height of waves would be of value to naval architects particularly in respect of stability, rolling and behaviour of the ship structure in a seaway.

CHAPTER 7

Observations of Ocean Currents and Ice

OCEAN CURRENT OBSERVATIONS

Introduction. Present knowledge of the general patterns of surface currents of the oceans has been largely derived from the systematically recorded observations made from ships on passage over more than a century. This knowledge is summarized in atlases, charts and Sailing Directions which have been prepared for the benefit of the navigator. However, much remains to be learnt about the currents, for instance about the general flow of waters that are remote from the main shipping routes, about local variations and intensities of currents and about the variabilities of currents through the course of the year and from one year to another. Consequently there is a continuing need for observations of surface currents. Moreover, as improved navigational aids allow increasing accuracy and frequency of observation, better and more detailed analyses of the currents will become possible, to the benefit not only of navigation but also of the various aspects of marine science.

Marine observers are reminded that towards the end of the Meteorological Logbook is a section reserved for recording the set and drift of currents observed during a voyage. Observations of ocean currents will be very welcome from any ship, whether reporting meteorological observations or not.

The method of observing currents. This consists of deriving the set and drift from the direction and distance respectively between a dead-reckoning (DR) position (obtained making due allowance for leeway) and the corresponding observed position. The current so deduced is the mean current affecting the ship over the distance between the fixes when the DR plot was started (the 'From' position) and finished (the 'To' position); it is assumed that this is the current at a depth of about half the ship's draught.

In calculating the DR position it is important that the course or courses should be corrected for leeway so that—as far as can be judged—the difference between DR and 'To' positions is due only to current and not to a combination of wind and current. The assessment of leeway can only be made by the mariner with his full knowledge of the vessel's performance and state of loading. If he has considerable doubt about the appropriate allowance for leeway—say because of gales during the period of the DR plot—an observation should not be recorded.

In general, observations should not be recorded when the derived value of current is likely to be unrepresentative or inaccurate. The following aspects concerning the representativeness and accuracy of observation should be borne in mind.

Representativeness of observations. The ideal observation of current would represent the purely non-tidal movement of water at a single point at a given time. Such an observation cannot be made in practice from a ship on passage but the departure from the ideal does not usually decrease the value of the results to a serious extent so long as certain limitations are heeded. An observation of current should not be made:

(a) where significant tidal streams are to be expected;

(b) when the distance between 'From' and 'To' positions exceeds about 400 miles;

(c) when the 'From' and 'To' positions are separated by more than a day's sailing—say by more than 25 hours, allowing for the ship's clocks being retarded;

(d) when there is reason to believe that the current changes significantly between the 'From' and 'To' positions, i.e. when passing from one system of currents to another, when passing a strait, a cape, or a current race;

(e) so as to overlap another observation: for example if there are successive fixes at positions A, B, C and D the overlapping observations from A to C and from B to D are not both required since they are not independent of each other; and

(f) when the accuracy of either the 'From' to 'To' fix is suspect.

Accuracy of observations. In effect the observed current is calculated from the difference between two vectors, i.e. the vector representing movement of the ship with respect to the water and the vector representing movement over the ocean bed. Since the difference is normally much smaller than either vector, the accuracies of values used to calculate these vectors crucially affect the accuracy of the inferred current.

The vector of movement with respect to the water depends on course (corrected for compass error and for leeway) and on distance run through the water. The application of compass error is of obvious importance; corrections for leeway have already been discussed. Distance run can be measured with high accuracy by means of some modern devices such as the electromagnetic log. If it is necessary to assess distance run from propeller revolutions the allowance for slip under the prevailing conditions (wind, sea, draught, state of ship's bottom) becomes important.

The accuracy of the vector of movement over the ocean bed is governed by the accuracy of the two fixes—for the 'From' and 'To' positions. Again some modern navigational aids can give especial accuracy, e.g. satellite navigator. Usefully accurate fixes may also be derived by a number of means, such as by observation of two or more stars at twilight, or by Loran or Omega navigational systems. A noon position based on forenoon sight and meridian altitude is not desirable for calculation of current if more accurate types of fix are available since the run between sights depends on a due appreciation of the current—the very element that is being sought.

Interval between 'From' and 'To' positions. As already stated the interval should not be too large; if the ship passes through two or more different currents then the observed (mean) value will be representative of none of the constituent currents. On the other hand the interval should not be too small since a given error in position fixing will produce an error in the calculated rate of current which is greater if the interval of time is short than if it is long. The combination of possibly inaccurate fix and short interval of time would make unreliable, for example, an observation based on a noon position, found from forenoon sight and meridian altitude, and a fix at approximately 1415, using the Sun and Venus. Indeed whenever use is made of a noon position based on forenoon and noon sights—with its inherent inaccuracies, however carefully compiled—the current should be calculated over the longest permissible interval of about 24 hours. At the other extreme, with both fixes highly accurate (say by satellite) an interval as short as three hours might be adequate.

ICE OBSERVATIONS

Under the International Convention for the Safety of Life at Sea, 1960, the master of every ship that meets with dangerous ice has the obligation to report this by all means available to ships in the vicinity and to the nearest coast radio station or signal station. (Details of the prescribed form of signal are given in Chapter 14.)

Apart from this primary obligation with regard to the immediate safety of life the mariner provides important contributions to the knowledge of ice at sea in two ways:

(a) By including in the routine synoptic weather report an ICE group or a plain-language report of ice whenever this is appropriate. This information is used especially for mapping the existing state and development of the ice as a basis for operational advice to mariners.

(b) By making entries in the section for ice reports towards the end of the Meteorological Logbook. This is a more detailed report than (a) above and its purpose is statistical rather than operational; the information is used for scientific investigations, including the preparation of maps showing average and extreme positions of ice edges. It should be borne in mind when no sea ice is encountered, though this is normally to be expected for the time of year and sea area, then a 'nil' report gives significant information and should be entered; such 'nil' reports can be as important as reports of ice sightings.

Some background knowledge of the physics and climatology of sea ice and icebergs is helpful to the observer. Such knowledge can be obtained from *Meteorology for Mariners*—Chapter 17 deals with formation and movement and Chapter 18 with distribution of sea ice and icebergs. Also *The Mariner's Handbook* (NP 100, Fourth Edition, published in 1973 by the Hydrographer of the Navy) contains a chapter on ice and gives a selection of photographs illustrating a number of ice terms. Some of these photographs are shown between pages 72 and 73.

To ensure consistency in the use of terms for ice these have been defined by the World Meteorological Organization (WMO) and published in 1970 in a glossary entitled *WMO sea-ice nomenclature*; the publication is comprehensively illustrated by photographs.

The terms of the *Nomenclature* with full definitions (incorporating the latest) amendments and arranged in alphabetical order are reproduced on pages 66 to 74. To assist the observer in selecting the term most appropriate for a certain state or process this alphabetical list is preceded by a section in which terms are arranged according to subject. Some brief explanatory remarks are included in this section but reference should be made to the alphabetical list for the full definition of any term.

ICE TERMS ARRANGED BY SUBJECT

1. ORIGIN OF ICE

   The terms **lake ice, river ice, sea ice** and **glacier ice** primarily refer to the origin of the ice, not its present location.

2. DEVELOPMENT

   (a) **New ice:** A general term for recently formed ice which includes *frazil ice, grease ice, slush* and *shuga.*

   (b) **Nilas:** A thin elastic crust of ice. Subdivisions are *dark nilas, light nilas.*

   (c) **Pancake ice:** Predominantly circular pieces of ice, with raised rims due to the pieces striking against one another.

   (d) **Young ice:** Ice in the transition stage between *nilas* and *first-year ice.* Subdivisions are *grey ice* and *grey-white ice.*

   (e) **First-year ice:** Sea ice of not more than one winter's growth, developing from *young ice.* Subdivisions are *thin first-year ice/white ice, medium first-year ice, thick first-year ice.*

   (f) **Old ice:** *Sea ice* which has survived at least one summer's melt. Subdivisions are *second-year ice, multi-year ice.*

3. FAST ICE

   Fast ice is *sea ice* which forms and remains fast along the coast, where it is attached to the shore, to an *ice wall*, to an *ice front*, between shoals or grounded *icebergs*. Some associated terms are *young coastal ice, icefoot, anchor ice, grounded ice, stranded ice* and *hummock.*

4. FLOATING ICE

   Floating ice is any form of ice floating in water and can include not only *sea ice* but also *river ice, lake ice* and *glacier ice (ice of land origin)*. Forms of floating ice are *pancake ice, floe, ice cake, floeberg, glacier berg, tabular berg, ice island, bergy bit* and *growler.*

5. PACK ICE

   Pack ice is used in a wide sense to include any area of *sea ice*, other than *fast ice*, no matter what form it takes or how it is disposed.

6. AREAL EXTENT AND AREAL DENSITY OF ICE

   Ice cover is used to express areal coverage of ice within a large geographical region, e.g. Baffin Bay, Barents Sea. **Concentration** is used to express areal density of ice in a given, comparatively small, area. The following descriptive terms imply different concentrations: *compact, consolidated, very close, open* and *very open pack ice; open water, ice-free.*

7. ARRANGEMENT OF ICE

   Ice field, ice massif, belt, tongue, strip, bight, ice jam, ice edge, ice boundary, iceberg tongue.

8. MOTION PROCESSES

   Diverging, compacting, shearing.

65

9. DEFORMATION PROCESSES

Fracturing, hummocking, ridging, rafting and *finger rafting* result from pressures exerted on the ice.
Weathering.

10. OPENINGS IN THE ICE

Fracture. See also *crack, tide crack, flaw* and *fracture zone.*
Lead. See also *shore lead, flaw lead.*
Polynya. See also *shore polynya, flaw polynya* and *recurring polynya.*

11. GENERAL CHARACTER OF ICE SURFACE

Deformed ice: Ice which has been squeezed together and in phases forced upwards (or downwards). Subdivisions are *rafted ice, ridged ice, hummocked ice.*
Level ice: *Sea ice* which is unaffected by deformation.

12. ICE-SURFACE FEATURES

Standing floe, ram, bare ice, snow-covered ice, sastrugi, snowdrift.

13. TERMS RELATED TO MELTING

Puddle, thaw holes, dried ice, rotten ice, flooded ice.

14. ICE OF LAND ORIGIN

Firn.
Glacier ice. See also *glacier, ice wall, ice stream, glacier tongue.*
Ice shelf. The seaward edge is termed an *ice front.*
Calved ice. See also *iceberg, ice island, bergy bit* and *growler.*

15. SKY AND AIR INDICATIONS

Water sky, ice blink, frost smoke.

16. TERMS RELATING TO SURFACE SHIPPING

Beset, ice-bound, nip, ice under pressure, difficult area, easy area, iceport.

17. TERMS RELATING TO SUBMARINE NAVIGATION

Ice canopy, friendly ice, hostile ice, bummock, ice keel, skylight.

ICE TERMS ARRANGED IN ALPHABETICAL ORDER

Aged ridge: *Ridge* which has undergone considerable weathering. These ridges are best described as undulations.

Anchor-ice: Submerged ice attached or anchored to the bottom, irrespective of the nature of its formation.

Bare ice: Ice without snow cover.

Belt: A large feature of *pack ice* arrangement, longer than it is wide, from 1 km to more than 100 km in width.

Bergy bit: A large piece of floating *glacier ice*, generally showing less than 5 m above sea level but more than 1 m and normally about 100–300 m² in area.

Bergy water: An area of freely navigable water with no *sea ice* present but in which *ice of land origin* is present.

Beset: Situation of a vessel surrounded by ice and unable to move.

Big floe: See *Floe*.

Bight: An extensive crescent-shaped indentation in the *ice edge*, formed by either wind or current.

Brash ice: Accumulations of *floating ice* made up of fragments not more than 2 m across, the wreckage of other forms of ice.

Bummock: From the point of view of the submariner, a downward projection from the underside of the *ice canopy*; the counterpart of a *hummock*.

Calving: The breaking away of a mass of ice from an *ice wall, ice front* or *iceberg*.

Close pack ice: *Pack ice* in which the *concentration* is 7/10 to 8/10 (6/8 to less than 7/8), composed of floes mostly in contact.

Compacted ice edge: Close, clear-cut *ice edge* compacted by wind or current; usually on the windward side of an area of *pack ice*.

Compacting: Pieces of *floating ice* are said to be compacting when they are subjected to a converging motion, which increases ice *concentration* and/or produces stresses which may result in ice deformation.

Compact pack ice: *Pack ice* in which the *concentration* is 10/10 (8/8) and no water is visible.

Concentration: The ratio expressed in tenths or oktas describing the mean areal density of ice in a given area.

Concentration boundary: A line approximating the transition between two areas of *pack ice* with distinctly different *concentrations*.

Consolidated pack ice: *Pack ice* in which the *concentration* is 10/10 (8/8) and the *floes* are frozen together.

Consolidated ridge: A *ridge* in which the base has frozen together.

Crack: Any *fracture* which has not parted.

Dark nilas: *Nilas* which is under 5 cm in thickness and is very dark in colour.

Deformed ice: A general term for ice which has been squeezed together and in places forced upwards (and downwards). Subdivisions are *rafted ice, ridged ice* and *hummocked ice*.

Difficult area: A general qualitative expression to indicate, in a relative manner, that the severity of ice conditions prevailing in an area is such that navigation in it is difficult.

Diffuse ice edge: Poorly defined *ice edge* limiting an area of dispersed ice; usually on the leeward side of an area of *pack ice*.

Diverging: *Ice fields* or *floes* in an area are subjected to diverging or dispersive motion, thus reducing ice *concentration* and/or relieving stresses in the ice.

Dried ice: *Sea ice* from the surface of which melt-water has disappeared after the formation of *cracks* and *thaw holes*. During the period of drying, the surface whitens.

Easy area: A general qualitative expression to indicate, in a relative manner, that ice conditions prevailing in an area are such that navigation in it is not difficult.

Fast ice: *Sea ice* which forms and remains fast along the coast, where it is attached to the shore, to an *ice wall*, to an *ice front*, between shoals or grounded *icebergs*. Vertical fluctuations may be observed during changes of sea level. Fast ice may be formed *in situ* from sea water or by freezing of *pack ice* of any age to the shore, and it may extend a few metres or several hundred kilometres from the coast. Fast ice may be more than one year old and may then be prefixed with the appropriate age category (*old, second-year* or *multi-year*). If it is thicker than about 2 m above sea level it is called an *ice shelf*.

Fast-ice boundary: The *ice boundary* at any given time between *fast ice* and *pack ice*.

Fast-ice edge: The demarcation at any given time between *fast ice* and *open water*.

Finger rafted ice: Type of *rafted ice* in which *floes* thrust 'fingers' alternately over and under the other.

Finger rafting: Type of rafting whereby interlocking thrusts are formed, each floe thrusting 'fingers' alternately over and under the other. Common in *nilas* and *grey ice*.

Firn: Old snow which has recrystallized into a dense material. Unlike snow, the particles are to some extent joined together; but, unlike ice, the air spaces in it still connect with each other.

First-year ice: *Sea ice* of not more than one winter's growth, developing from *young ice*: thickness 30 cm–2 m. May be subdivided into *thin first-year ice/white ice, medium first-year ice* and *thick first-year ice*.

Flaw: A narrow separation zone between *pack ice* and *fast ice*, where the pieces of ice are in chaotic state; it forms when pack ice shears under the effect of a strong wind or current along the *fast ice boundary* (cf. *shearing*).

Flaw lead: A passage-way between *pack ice* and *fast ice* which is navigable by surface vessels.

Flaw polynya: A *polynya* between *pack ice* and *fast ice*.

Floating ice: Any form of ice found floating in water. The principal kinds of floating ice are *lake ice, river ice*, and *sea ice*, which form by the freezing of water at the surface, and *glacier ice* (*ice of land origin*) formed on land or in an *ice shelf*. The concept includes ice that is stranded or grounded.

Floe: Any relatively flat piece of *sea ice* 20 m or more across. Floes are sub-divided according to horizontal extent as follows:
GIANT: Over 5·4 n. mile across.
VAST: 1·1–5·4 n. mile across.
BIG: 500–2000 m across.
MEDIUM: 100–500 m across.
SMALL: 20–100 m across.

Floeberg: A massive piece of *sea ice* composed of a *hummock*, or a group of *hummocks*, frozen together and separated from any ice surroundings. It may float up to 5 m above sea level.

Flooded ice: *Sea ice* which has been flooded by melt-water or river water and is heavily loaded by water and wet snow.

Fracture: Any break or rupture through *very close pack ice, compact pack ice, consolidated pack ice, fast ice*, or a single *floe* resulting from deformation processes. Fractures may contain *brash ice* and/or be covered with *nilas* and/or *young ice*. Length may vary from a few metres to many nautical miles.

Fracture zone: An area which has a great number of fractures.

Fracturing: Pressure process whereby ice is permanently deformed, and rupture occurs. Most commonly used to describe breaking across *very close pack ice, compact pack ice* and *consolidated pack ice*.

Frazil ice: Fine spicules or plates of ice, suspended in water.

Friendly ice: From the point of view of the submariner, an *ice canopy* containing many large *skylights* or other features which permit a submarine to surface. There must be more than 10 such features per 30 n. mile (56 km) along the submarine's track.

Frost smoke: Fog-like clouds due to contact of cold air with relatively warm water, which can appear over openings in the ice, or leeward of the *ice edge*, and which may persist while ice is forming.

Giant floe: See *Floe*.

Glacier: A mass of snow and ice continuously moving from higher to lower ground or, if afloat, continuously spreading. The principal forms of glacier are: inland ice sheets, *ice shelves, ice streams*, ice caps, ice piedmonts, cirque glaciers and various types of mountain (valley) glaciers.

Glacier berg: An irregularly shaped *iceberg*.

Glacier ice: Ice in, or originating from, a *glacier*, whether on land or floating on the sea as *icebergs, bergy bits* or *growlers*.

Glacier tongue: Projecting seaward extension of a *glacier*, usually afloat. In the Antarctic glacier tongues may extend over many tens of nautical miles.

Grease ice: A later stage of freezing than *frazil ice* when the crystals have coagulated to form a soupy layer on the surface. Grease ice reflects little light, giving the sea a matt appearance.

Grey ice: *Young ice* 10–15 cm thick. Less elastic than *nilas* and breaks on swell. Usually rafts under pressure.

Grey-white ice: *Young ice* 15–30 cm thick. Under pressure more likely to ridge than to raft.

Grounded hummock: Hummocked *grounded ice* formation. There are single grounded *hummocks* and lines (or chains) of grounded *hummocks*.

Grounded ice: *Floating ice* which is aground in shoal water (cf. *stranded ice*).

Growler: Smaller piece of ice than a *bergy bit* or *floeberg*, often transparent but appearing green or almost black in colour, extending less than 1 m above the sea surface and normally occupying an area of about 20 m².

Hostile ice: From the point of view of the submariner, an *ice canopy* containing no large *skylights* or other features which permit a submarine to surface.

Hummock: A hillock of broken ice which has been forced upwards by pressure. May be fresh or weathered. The submerged volume of broken ice under the hummock, forced downwards by pressure, is termed a *bummock*.

Hummocked ice: *Sea ice* piled haphazardly one piece over another to form an uneven surface. When weathered, has the appearance of smooth hillocks.

Hummocking: The pressure process by which *sea ice* is forced into *hummocks*. When the floes rotate in the process it is termed screwing.

Iceberg: A massive piece of ice greatly varying in shape, more than 5 m above sea level, which has broken away from a *glacier*, and which may be afloat or aground. Icebergs may be described as *tabular*, dome-shaped, sloping, pinnacled, weathered or *glacier bergs*.

Iceberg tongue: A major accumulation of *icebergs* projecting from the coast, held in place by grounding and joined together by *fast ice*.

Ice blink: A whitish glare on low clouds above an accumulation of distant ice.

Ice-bound: A harbour, inlet etc. is said to be ice-bound when navigation by ships is prevented on account of ice, except possibly with the assistance of an ice-breaker.

Ice boundary: The demarcation at any given time between *fast ice* and *pack ice* or between areas of *pack ice* of different *concentrations* (cf. *ice edge*).

Ice breccia: Ice pieces of different age frozen together.

Ice cake: Any relatively flat piece of *sea ice* less than 20 m across. See *Small ice cake*.

Ice canopy: *Pack ice* from the point of view of the submariner.

Ice cover: The ratio of an area of ice of any concentration to the total area of sea surface within some large geographic locale; this locale may be global, hemispheric, or prescribed by a specific oceanographic entity such as Baffin Bay or the Barents Sea.

Ice edge: The demarcation at any given time between the open sea and *sea ice* of any kind, whether fast or drifting. It may be termed *compacted* or *diffuse* (cf. *ice boundary*).

Ice field: Area of *pack ice* consisting of any size of floes, which is greater than 5·4 n. mile across (cf. *ice patch*). See *Large, Medium, Small ice field*.

Icefoot: A narrow fringe of ice attached to the coast, unmoved by tides and remaining after the *fast ice* has moved away.

Ice-free: No ice present. If ice of any kind is present this term should not be used.

Ice front: The vertical cliff forming the seaward face of an *ice shelf* or other floating *glacier* varying in height from 2 to 50 m or more above sea level (cf. *ice wall*).

Ice island: A large piece of floating ice about 5 m above sea level, which has broken away from an Arctic ice shelf, having a thickness of 30–50 m and an area of from a few thousand square metres to 150 n. mile^2 or more, and usually characterized by a regularly undulating surface which gives it a ribbed appearance from the air.

Ice jam: An accumulation of broken *river ice* or *sea ice* caught in a narrow channel.

Ice keel: From the point of view of the submariner, a downward-projecting ridge on the underside of the *ice canopy*, the counterpart of a ridge. Ice keels may extend as much as 50 m below sea level.

Ice Limit: Climatological term referring to the extreme minimum or extreme maximum extent of the *ice edge* in any given month or period based on observations over a number of years. Term should be preceded by minimum or maximum (cf. *mean ice edge*).

Ice massif: A concentration of *sea ice* covering hundreds of nautical miles square, which is found in the same region every summer.

Ice of land origin: Ice formed on land or in an *ice shelf*, found floating in water. The concept includes ice that is stranded or grounded.

Ice patch: An area of *pack ice* less than 5·4 n. mile across.

Ice port: An embayment in an *ice front*, often of a temporary nature, where ships can moor alongside and unload directly on to the ice shelf.

Ice rind: A brittle shiny crust of ice formed on a quiet surface by direct freezing or from *grease ice*, usually in water of low salinity. Thickness to about 5 cm. Easily broken by wind or swell, commonly breaking in rectangular pieces.

Ice shelf: A floating ice sheet of considerable thickness showing 2–50 m or more above sea level, attached to the coast. Usually of great horizontal extent and with a level or gently undulating surface. Nourished by annual snow accumulation and often also by the seaward extension of land *glaciers*. Limited areas may be aground. The seaward edge is termed an *ice front*.

Ice stream: Part of an inland ice sheet in which the ice flows more rapidly and not necessarily in the same direction as the surrounding ice. The margins

70

are sometimes clearly marked by a change in direction of the surface slope but may be indistinct.

Ice under pressure: Ice in which deformation processes are actively occurring and hence a potential impediment or danger to shipping.

Ice wall: An ice cliff forming the seaward margin of a *glacier* which is not afloat. An ice wall is aground, the rock basement being at or below sea level (cf. *ice front*).

Lake ice: Ice formed on a lake, regardless of observed location.

Large fracture: More than 500 m wide.

Large ice field: An *ice field* over 11 n. mile across.

Lead: Any *fracture* or passage-way through *sea ice* which is navigable by surface vessels.

Level ice: *Sea ice* which is unaffected by deformation.

Light nilas: *Nilas* which is more than 5 cm in thickness and rather lighter in colour than *dark nilas*.

Mean ice edge: Average position of the *ice edge* in any given month or period based on observations over a number of years. Other terms which may be used are mean maximum ice edge and mean minimum ice edge (cf. *ice limit*).

Medium first-year ice: *First-year ice* 70–120 cm thick.

Medium floe: See *Floe*.

Medium fracture: 200–500 m wide.

Medium ice field: An *ice field* 8–11 n. mile across.

Multi-year ice: *Old ice* up to 3 m or more thick which has survived at least two summers' melt. *Hummocks* even smoother than in second-year ice, and the ice is almost salt-free. Colour, where bare, is usually blue. Melt pattern consists of large interconnecting irregular *puddles* and a well-developed drainage system.

New ice: A general term for recently formed ice which includes *frazil ice, grease ice, slush* and *shuga*. These types of ice are composed of ice crystals which are only weakly frozen together (if at all) and have a definite form only while they are afloat.

New ridge: *Ridge* newly formed with sharp peaks and slope of sides usually 40°. Fragments are visible from the air at low altitude.

Nilas: A thin elastic crust of ice, easily bending on waves and swell and under pressure, thrusting in a pattern of interlocking 'fingers' (*finger rafting*). Has a matt surface and is up to 10 cm in thickness. May be subdivided into *dark nilas* and *light nilas*.

Nip: Ice is said to nip when it forcibly presses against a ship. A vessel so caught, though undamaged, is said to have been nipped.

Old ice: *Sea ice* which has survived at least one summer's melt. Most topographic features are smoother than on *first-year ice*. May be subdivided into *second-year ice* and *multi-year ice*.

Open pack ice: *Pack ice* in which the ice *concentration* is 4/10 to 6/10 (3/8 to less than 6/8), with many *leads* and *polynyas*, and the *floes* are generally not in contact with one another.

Open water: A large area of freely navigable water in which *sea ice* is present in *concentrations* less than 1/10 (1/8). There may be *ice of land origin* present, although the total concentration of all ice shall not exceed 1/10 (1/8).

Pack ice: Term used in a wide sense to include any area of *sea ice*, other than *fast ice*, no matter what form it takes or how it is disposed.

Pancake ice: Predominantly circular pieces of ice from 30 cm to 3 m in diameter, and up to about 10 cm in thickness, with raised rims due to the pieces striking against one another. It may be formed on a slight swell from *grease ice, shuga* or *slush* or as a result of the breaking of *ice rind, nilas* or, under severe conditions of swell or waves, of *grey ice*. It also sometimes forms at some depth, at an interface between water bodies of different physical characteristics, from where it floats to the surface; its appearance may rapidly cover wide areas of water.

Polynya: Any non-linear shaped opening enclosed in ice. Polynyas may contain *brash ice* and/or be covered with *new ice, nilas* or *young ice*; submariners refer to these as *skylights*. Sometimes the polynya is limited on one side by the coast and is called a *shore polynya* or by *fast ice* and is called a *flaw polynya*. If it recurs in the same position every year, it is called a *recurring polynya*.

Puddle: An accumulation on ice of melt-water, mainly due to melting snow, but in the more advanced stages also to the melting of ice. Initial stage consists of patches of melted snow.

Rafted ice: Type of *deformed ice* formed by one piece of ice overriding another (cf. *finger rafting*).

Rafting: Pressure processes whereby one piece of ice overrides another. Most common in *new* and *young ice* (cf. *finger rafting*).

Ram: An underwater ice projection from an *ice wall, ice front, iceberg* or *floe*. Its formation is usually due to a more intensive melting and erosion of the unsubmerged part.

Recurring polynya: A *polynya* which recurs in the same position every year.

Ridge: A line or wall of broken ice forced up by pressure. May be fresh or weathered. The submerged volume of broken ice under a ridge, forced downwards by pressure, is termed an *ice keel*.

Ridged ice: Ice piled haphazardly one piece over another in the form of ridges or walls. Usually found in first-year ice (cf. *ridging*).

Ridged-ice zone: An area in which much *ridged ice* with similar characteristics has formed.

Ridging: The pressure process by which *sea ice* is forced into *ridges*.

River ice: Ice formed on a river, regardless of observed location.

Rotten ice: *Sea ice* which has become honeycombed and which is in an advanced state of disintegration.

Sastrugi: Sharp, irregular ridges formed on a snow surface by wind erosion and deposition. On mobile *floating ice* the ridges are parallel to the direction of the prevailing wind at the time they were formed.

Sea ice: Any form of ice found at sea which has originated from the freezing of sea water.

Second-year ice: *Old ice* which has survived only one summer's melt. Because it is thicker and less dense than *first-year ice*, it stands higher out of the water. In contrast to *multi-year ice*, summer melting produces a regular pattern of numerous small *puddles*. Bare patches and puddles are usually greenish-blue.

Shearing: An area of *pack ice* is subject to shear when the ice motion varies significantly in the direction normal to the motion, subjecting the ice to rotational forces. These forces may result in phenomena similar to a *flaw* (q.v.).

Shore lead: A *lead* between *pack ice* and the shore or between *pack ice* and an *ice front*.

Photograph by Dr T. Armstrong, Scott Polar Research Institute

1. Grease Ice

I

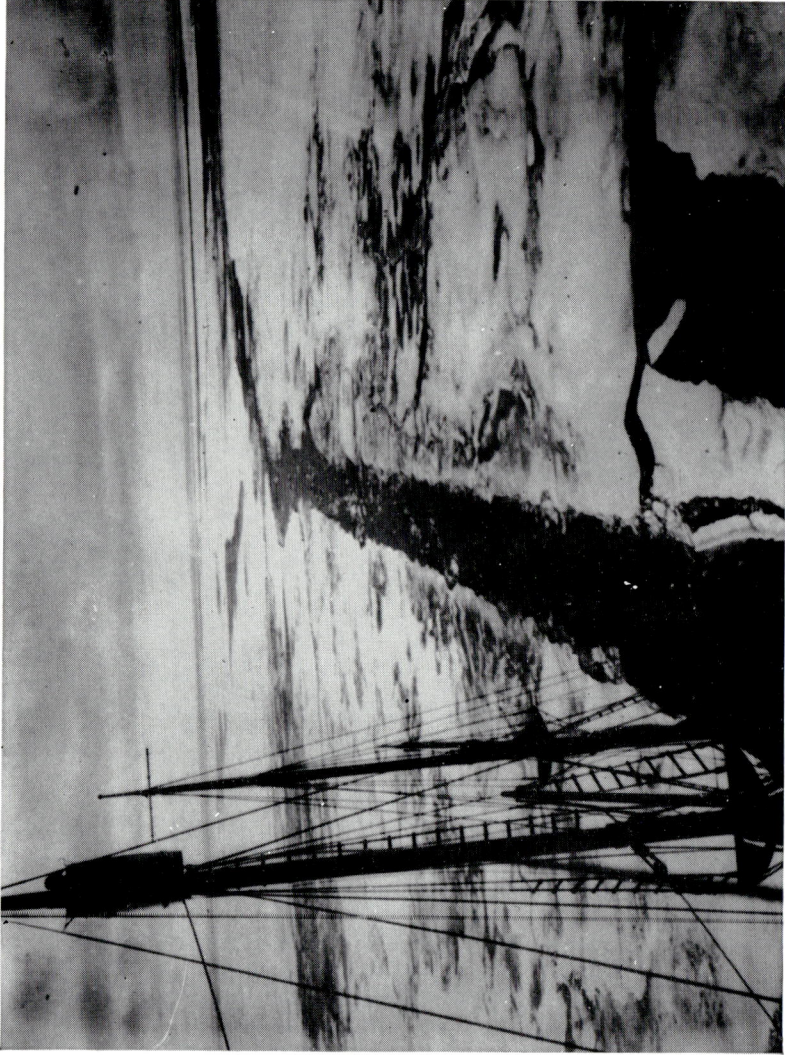

2. Light Nilas

Photograph by Dr Swithinbank, Scott Polar Research Institute

3. Very open Pack Ice

I*

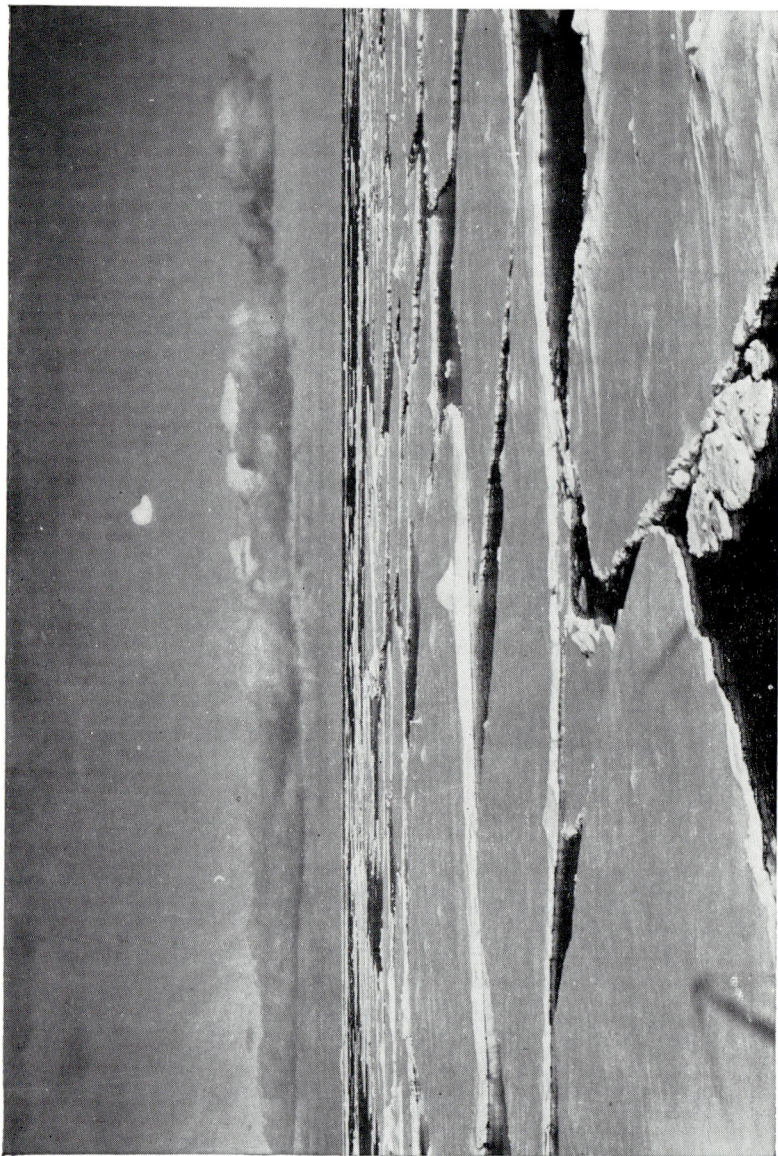

4. Open Pack Ice

5. Very close Pack Ice

Photograph by Dr Swithinbank, Scott Polar Research Institute

6. Ice Edge

7. Lead

Photograph by J. F. Hurley

8. Hummocked Ice

9. Growler

Photograph by G. Murray-Levick

10. Bergy Bit

11. Glacier Berg

12. Tabular Berg

Shore polynya: A *polynya* between *pack ice* and the coast or between *pack ice* and an *ice front*.

Shuga: An accumulation of spongy white ice lumps, a few centimetres across; they are formed from *grease ice* or *slush* and sometimes from *anchor-ice* rising to the surface.

Skylight: From the point of view of the submariner, thin places in the *ice canopy*, usually less than 1 m thick and appearing from below as relatively light, translucent patches in dark surroundings. The under-surface of a skylight is normally flat. Skylights are called large if big enough for a submarine to attempt to surface through them (120 m), or small if not.

Slush: Snow which is saturated and mixed with water on land or ice surfaces, or as a viscous floating mass in water after a heavy snowfall.

Small floe: See *Floe*.

Small fracture: 50–200 m wide.

Small ice cake: An ice cake less than 2 m across.

Small ice field: An *ice field* 5·4–8 n. mile across.

Snow-covered ice: Ice covered with snow.

Snowdrift: An accumulation of wind-blown snow deposited in the lee of obstructions or heaped by wind eddies. A crescent-shaped snowdrift, with ends pointing downwind, is known as a snow barchan.

Standing floe: A separate *floe* standing vertically or inclined and enclosed by rather smooth ice.

Stranded ice: Ice which has been floating and has been deposited on the shore by retreating high water.

Strip: Long narrow area of *pack ice*, about 0·5 n. mile or less in width, usually composed of small fragments detached from the main mass of ice, and run together under the influence of wind, swell or current.

Tabular berg: A flat-topped *iceberg*. Most tabular bergs form by *calving* from an *ice shelf* and show horizontal banding (cf. *ice island*).

Thaw holes: Vertical holes in sea ice formed when surface *puddles* melt through to the underlying water.

Thick first-year ice: *First-year ice* over 120 cm thick.

Thin first-year ice/white ice: *First-year ice* 30–70 cm thick.

Tide crack: Crack at the line of junction between an immovable *icefoot* or *ice wall* and *fast ice*, the latter subject to rise and fall of the tide.

Tongue: A projection of the ice edge up to several nautical miles in length, caused by wind or current.

Vast floe: See *Floe*.

Very close pack ice: *Pack ice* in which the *concentration* is 9/10 to less than 10/10 (7/8 to less than 8/8).

Very open pack ice: *Pack ice* in which the *concentration* is 1/10 to 3/10 (1/8 to less than 3/8) and water preponderates over ice.

Very small fracture: 0–50 m wide.

Very weathered ridge: *Ridge* with tops very rounded, slopes of sides usually 20°–30°.

Water sky: Dark streaks on the underside of low clouds, indicating the presence of water features in the vicinity of *sea ice*.

Weathered ridge: *Ridge* with peaks slightly rounded and slope of sides usually 30°–40°. Individual fragments are not discernible.

Weathering: Processes of ablation and accumulation which gradually eliminate irregularities in an ice surface.

K

73

White ice: See *Thin first-year ice*.

Young coastal ice: The initial stage of *fast ice* formation consisting of *nilas* or *young ice*, its width varying from a few metres up to 100–200 m from the shoreline.

Young ice: Ice in the transition stage between *nilas* and *first-year ice*, 10–30 cm in thickness. May be subdivided into *grey ice* and *grey-white ice*.

Part III   Phenomena

CHAPTER 8

The Observation of Phenomena

Introduction. The seaman has unusual opportunities for observing natural phenomena of all kinds. This can be made an interesting hobby, and the observer may be lucky enough, sooner or later, to make a rare, or even unique, observation, which if carefully observed and recorded, will contribute to scientific knowledge. The comparative frequency or rarity of certain phenomena is indicated in this and the four following chapters, as far as our present knowledge goes. Phenomena of unknown origin are occasionally seen at sea and these should be carefully observed and recorded.

It is however not only the rare observations which are of value. All meteorological phenomena, whether optical or general, are directly related to the state of the atmosphere and weather prevailing at the time, and their recording in the Remarks Column of the Meteorological Log or in the space provided for Additional Remarks, helps to complete the information given by routine observations. Also there is probably a good deal to be learnt yet about many of the more common phenomena, including their frequency and geographical distribution, for which purpose it is obvious that all observations made in any part of the world should be put on record.

Hints are given in these chapters on the observations or measurements which are necessary if the phenomenon is to be correctly identified. Observations are much more valuable if accompanied by drawings or sketches, in black and white or colour, or by photographs. If there is not room in the log, the observations and sketches can be attached to it.

The more interesting and unusual observations and illustrations are published in *The Marine Observer*. Notes and sketches on phenomena which are outside the scope of the Meteorological Office are always sent on to the relevant authority for examination and comment.

Methods of observation. Some optical phenomena such as coronae and iridescent cloud appear to be very near the sun or moon. Those near the sun may not be seen at all unless the eyes are shaded from direct sunlight. Apart from this, optical phenomena such as halos, coronae etc., viewed in the day-time, when the sky is often very bright, are more easily seen if the amount of light entering the eye is reduced, and sometimes a very faint halo etc., can only be seen if this is done. The sky may be viewed through neutral-tinted glass of a light tone, such as the lightest of those on a sextant, or the reflection from black glass may be used, if available, or from a piece of ordinary glass painted on one side with black enamel or backed with black paper. If a pair of ordinary sun-glasses of suitable colour is available, this is the best method of all. Yellow-brown, not too deep, has been found to be very satisfactory. Glasses of this colour have the power of slightly increasing contrast, so as to show distant land more distinctly on a misty day. The natural colour of any phenomenon is, of course, modified by these. The same methods also give a better view of clouds, of the details in a bright cloud mass, or of the very faint extensions, near the limit of visibility, of cloud in a blue sky.

There is a useful tip for seeing any very faint light at night, which is near or just beyond the limit of direct visibility. Do not look directly at the object, or where you suspect it to be, but fix the attention on a point a little way above, below, or to the side of it. Then view the spot 'out of the corner of the eye'. Light will thus be seen that would otherwise be invisible or, if it is directly visible, it will appear brighter by this process of 'averted' or 'oblique' vision. This applies to very faint light of every sort, whether concentrated in a point or diffused, such as faint terrestrial lights, faint stars, comets' tails, all zodiacal light phenomena and the fainter parts of aurorae.

In the case of phenomena of considerable duration, it is best to make notes of the various appearances as they are seen to come into view, or of other changes, carefully recording the times throughout the progress of the phenomenon. This is preferable to trusting to the memory afterwards. Rough sketches can also be made at the time and subsequently worked up into finished drawings or sketches. If colour is to be used, notes of the various colours should be made at the time.

It is desirable that observations be accompanied by sketches and/or photographs, whenever possible. These will often show detail that cannot be put into words. Sketches may be made either in ink or in pencil; in some cases delicacy of shading or fine detail is better rendered in pencil. When chosen for reproduction in *The Marine Observer*, the sketches are redrawn on tracing linen to enable engraved blocks to be made for the printer. Although sketches in colour cannot be so reproduced in *The Marine Observer* they may be of value in amplifying the detail given in the written observation.

An accurate account of the size and relative positions of the main features of what is seen is the prime requirement. Angular measurements are necessary in many cases for the identification of the phenomenon, as explained in the subsequent chapters. These are best incorporated in the written observation, unless the accompanying sketch is purely a diagrammatic one.

Phenomena such as haloes, rainbows and waterspouts may be photographed, giving a sufficiently short exposure, such as would best show cloud detail. The best results, particularly in the case of coloured objects, can only be got by the use of colour film or panchromatic film of suitable speed with a suitable colour filter over the lens. The same remarks apply to mirage, which has very rarely been photographed, though there appears to be no reason why satisfactory results should not be obtained.

OBSERVATIONS BY RADAR

With radar sets working on 3 cm or 10 cm, having a suitable form of presentation (e.g. PPI), echoes are obtained from rain up to distances of 40 n. mile or more. In this way showers, fronts and thunderstorms may be located and warning given of their approach. Echoes from cloud have been reported, but these are probably due to rain or drizzle within the cloud and not to the cloud particles themselves.

Objects at ground level or sea level are normally visible on the radar screen at distances a little beyond the visible horizon, owing to refraction. In certain conditions, however, much greater ranges are obtained. This occurs most frequently over the sea, and is due to a temperature inversion near the surface and/or a fall of humidity with height which causes reflection or abnormal refraction of the rays.

The reverse effect, i.e. a smaller degree of refraction than is usual—or sub-refraction—can occur owing to a very pronounced temperature lapse rate and/or an increase of humidity with height. Sub-refraction however is neither a very marked nor frequent phenomenon.

Ordinary meteorological fronts are not a major cause of abnormal radar ranges. Due to absorption of the radio energy, very heavy rain may tend to mask a radar target behind the rain area; this effect is unlikely to be significant on a wave-length of 10 cm but it may become important at shorter wave-lengths.

The use of radar as a means of detection of ice should be borne in mind. In normal meteorological conditions, echoes from most bergs may be detected at a useful range, but in certain meteorological conditions sub-refraction may occur and normal detection ranges be appreciably reduced. It has also been found that at times, even under favourable conditions, a very poor echo has been obtained from quite a large berg, the inclination of the slope presented to the observer apparently having an effect upon its reflecting properties, which is of as much account as the length or height of the berg. Bergy bits, growlers or pieces of pack ice, especially if smoothed by weathering, may pass undetected in strong sea clutter, even if they are large enough to sink or damage ships. On the other hand, in conditions where sea clutter is well marked, the cessation of such echoes may indicate the presence of pack ice.

Radar can therefore only be considered as an additional aid to the navigator in the detection of ice but it must be clearly understood that an absence of indication on the screen does not necessarily mean the absence of dangerous ice in the nieghbourhood of the vessel.

CHAPTER 9
Astronomical Phenomena
ECLIPSES

Partial eclipses of the sun or moon provide interesting spectacles but afford no opportunity for the seaman to make observations of particular value. Little diminution in sunlight is perceived until more than half the sun's disc is covered by the moon. An appreciable fall of temperature occurs during a large partial eclipse of the sun.

A *total* eclipse of the *sun* is perhaps the grandest of all natural phenomena. While almost of annual occurrence, its visibility on any occasion is confined to a very small area, along a line usually less than 100 n. mile wide, so that in any fixed place it is in general very rare. The duration of the total phase is very short, usually from a few seconds up to about two minutes, though in very exceptional circumstances it may be considerably more, the possible maxima being nearly eight minutes. During totality the fall of temperature is marked; often the wind changes or springs up, if previously calm. The sky darkens and has a peculiar appearance, often with lurid cloud colours. During totality the bright planets and the brighter stars may be seen.

Very occasionally a ship at sea or in harbour may be on the line of totality and several of such observations have been received in the last 50 years. The seaman fortunate enough to witness such an eclipse should endeavour to record all that he sees in as full detail as possible. There is so much to see in such a short time that it is desirable for several persons to observe in company. At the instant the moon finally covers the round body of the sun normally seen, the solar corona will spring into view. This is an irregularly extended atmosphere of the sun, pearly-white in colour, giving about half as much light as the full moon. It has a definite shape which varies according to the position of the year of observation in the 11-year cycle of solar activity (see under Sunspots). Near the time of maximum activity the corona is disposed fairly equally round the sun, with a definite structure of rays and bands, and sometimes curved forms like flower petals. Near the time of minimum activity the corona shows much less structural detail and the form is quite different. A wide band, more or less parallel sided, stretches outward from the equatorial region of the sun, one on each side of the sun, and these bands may extend a long distance, up to two or more solar diameters. At this time the polar regions usually show only a few short rays of coronal light. In the intermediate years of the solar cycle, the corona assumes forms intermediate between those described above.

Owing to the short duration of total solar eclipses and their comparative rarity, the total time for which the corona has been seen in the last 150 years is probably about two hours. Its exact form on any particular occasion is unpredictable. Marine observers can therefore make observations of real scientific value if the form, extent and detail of the corona is carefully noted and sketched. As the fainter extensions of the corona are best seen with the unaided eye and the structural detail is best seen with binoculars or a small telescope, it is best, especially when the duration of totality is short, to have two observers, each working in one of these different ways.

78

One or more of the great rose-red eruptions of hydrogen and calcium gas from the sun, known as prominences, may be seen adjacent to the moon's limb without optical assistance, especially if the sun is near its state of maximum activity. Unlike the corona, these may be seen in full sunlight on any day, by astronomers using special apparatus. Other features of a total eclipse on which attention may be concentrated are (a) meteorological effects, (b) the changing colour effects of sky and cloud and the rapid onrush of the moon's shadow through the air as the total phase begins, (c) the visibility of planets and stars.

The total phase of a *lunar* eclipse generally lasts a considerable time, sometimes for nearly two hours; the exact duration depends on how centrally the moon passes through the earth's shadow. The totally eclipsed moon usually remains visible, appearing in some shade of red or copper. Careful observation of this colour, and its changes, if any, during the total phase are of value. A general statement of the degree of brightness of the totally eclipsed moon should also be given, noting how far its surface markings remain visible. The totally eclipsed moon receives reddish sunlight by refraction through the section of the earth's atmosphere in profile to the moon at the time, and the amount and colour of the refracted light vary according to the cloudiness and other meteorological conditions in this part of the atmosphere. When fine dust in sufficient quantity is suspended in the air after a big volcanic eruption, the moon may almost, or even completely, disappear from sight during total eclipse. Such an observation should be carefully recorded, with all relevant detail.

COMETS

Comets are members of the Solar System, moving in elliptical orbits, in most cases so enormously elongated that the period of revolution round the sun may be hundreds or even thousands of years. A few return in a comparatively short time, one of these being the well-known Halley's Comet, with a period of about 76 years, last seen in 1910.

Comets are much less dense than planets, and consist of a loose aggregation of widely separated small solid bodies, ranging from the size of a grain of sand to that of small stones, probably with an admixture of larger pieces. The diameter of this collection is usually only a few hundred miles, but may be several thousand. Comets are only seen in that part of their orbit near the sun, when they shine partly by reflected light but mainly by the vaporizing of the material of the comet by the sun's heat. An interesting feature of a comet is its tail, which is only formed when the comet is relatively near the sun. This consists of dust and gases ejected from the head, probably by light pressure and electrical repulsion. The tail of a large comet may be many millions of miles in actual length. The apparent length may be anything from a degree or two to 60° or 80° or more. The direction of the tail is from the comet's head away from the sun. This direction bears no relation to the direction of the movement of the head of the comet in its orbit. The tail of a comet, unlike the transitory trail of a meteor, therefore does not show the direction in which the comet has travelled.

Most comets never become bright enough to be seen without telescopic aid and some never develop tails, but a bright comet is a magnificent naked-eye spectacle. There should be no confusion between the appearance of a comet and a meteor. A meteor is only seen for a few seconds as it travels more or less rapidly over its apparent path in the sky. A comet remains apparently fixed among the stars and sets with them in due course. It has a continuous movement relative to the stars, but in most cases this can only be seen in a naked-eye or binocular observation by comparing its position on successive nights. The period of naked-eye visibility of a comet may be anything from a few days to a number of weeks. It finally becomes invisible by either getting too faint, or passing into the daylight region of the sky or changing in declination so as to sink below the horizon.

Astronomers measure the position of the head of a comet relative to stars near it in the field of view of a telescope, or large-scale photographs may be taken. From a minimum of three such observations on successive nights, the comet's orbit in space and its subsequent apparent track in the sky can be computed. Angular distances of the comet from two or three bright stars, measured by sextant, are not sufficiently accurate for this purpose, but serve to identify the object and help in making an accurate sketch of the comet and its tail in relation to the stars. It may occasionally happen that more than one naked-eye comet is visible at the same time.

Valuable observations of a naked-eye comet may be made at sea, and it may happen that some interesting feature is seen which would not otherwise be put on record, if conditions of daylight or cloud make observations impossible in other parts of the world at that particular time. The brightness of the head and the form and length of the tail may sometimes change appreciably from night to night. The brightness of the head is estimated by comparison with that of neighbouring stars or planets, as described under Novae below. The altitude of the comet's head should be given, as part of this observation, also notes on the state of the sky, such as whether thin cloud, haze, twilight or moonlight is present. Careful sketches of the form and length of the tail are valuable and should include details of the structure of the tail, if any are seen, stating whether the observation was made with the unaided eye, or with binoculars. The end of the tail usually fades very gradually into the dark sky and the method of averted vision (see page 76) can be used to see it as far as possible; binoculars will not show the fainter extension. It is of special importance to record any tails, other than the main one, which may be visible; these are normally on the same side of the head as the main tail, making various angles with it, and they are usually narrower and fainter than the main tail. On rare occasions a short tail pointing towards the sun may be seen, i.e. in a direction opposite to that of the main tail. If the comet shows any peculiarity of colour this should be noted.

THE ZODIACAL LIGHT AND ASSOCIATED PHENOMENA

The zodiacal light. This is observed as the cone-shaped extremity of an elongated ellipse of soft whitish light which extends from the sun as centre, extending above the westerly horizon in the evening or the easterly horizon in the morning. The best time for observation is just after the last traces of twilight have disappeared in the evening, or just before the first traces appear in the morning. The light retains its apparent place among the stars and gradually sets or rises with them. It is more brilliant in the tropics, but is very conspicuous even in temperate latitudes, if observed away from the glare of large towns.

The axis of the light lies in the zodiac, very nearly but not quite in the plane of the ecliptic. In tropical latitudes, where the ecliptic makes a large angle with the horizon at all times of the year, the light may be seen well on any clear night or morning in all months. In the temperate latitudes of the northern hemisphere it is best seen in the evenings of January to March and in the mornings of September to November.

The light is pearly and homogeneous and differs markedly in quality from that of the Milky Way, the brightest part of which it may considerably exceed in luminosity. Its luminosity decreases with altitude above the horizon, since its brightness is greater the nearer the observed point is to the sun's position below the horizon. It appears, however, to fall off in brightness near the horizon on account of the greater thickness of the atmosphere its light has to traverse. At any altitude the axis of the light is brighter than its lateral parts. In northern temperate latitudes the edge of the cone towards the north in azimuth is less well-defined than that towards the south and tends to spread northwards near the horizon.

The zodiacal light is believed to be a cosmic phenomenon, due to the reflection of the sun's light from dust or gaseous matter, extending outwards to a point somewhat beyond the earth's orbit. There is much that is not known about this phenomenon and new observations from all latitudes will be of real value. Any features of interest should be noted, such as the colour of the light and any irregularity of form or light distribution. Observations of its brightness will be of value, as it is not yet known whether this is constant on successive nights or in different years. Apparent changes of brightness often occur since the night sky is not always equally transparent. The presence of a bright planet, especially Venus, in the region of the light dims it considerably. Estimates of brightness should be made on moonless nights, after all twilight has disappeared, by comparing the light with that of the Milky Way, preferably at about the same altitude. The position of the Milky Way should be specified, as this varies markedly in brightness in different parts of the sky. Thus the light on a given night might be estimated to be twice as bright as the Milky Way in Cygnus.

Observations of the precise position of the light, about which there is still some uncertainty, may be made by a careful sketch of the cone showing the position of specified stars, either within, on the edge of, or outside the cone.

Zodiacal band and Gegenschein. Joining the apexes of the cones of the morning and evening zodiacal lights is an extremely faint luminous band, a few degrees wide, lying along or nearly along the ecliptic, called the zodiacal band. On this band, at a point very nearly or exactly 180° from the sun's position in the ecliptic, is a somewhat brighter and larger but ill-defined patch, 10° or more in diameter, known by the German name 'Gegenschein'. This therefore is due south (in the northern hemisphere) at midnight, local time. These phenomena may be observed in temperate latitudes on the clearest moonless nights when at sufficient altitude; they are somewhat brighter in the tropics, on account of the greater altitude of the ecliptic. Further observations of these phenomena are much desired, especially from tropical localities. The track and width of the band, and the size, shape and position of the Gegenschein should be noted, together with variations of brilliancy and any special features seen, but the observation will be found difficult even to keen eyesight. The Gegenschein is usually invisible for the few nights on which it is projected upon the Milky Way in its annual journey round the ecliptic.

NOVAE

Sometimes, quite unpredictably, a small star, usually such that a telescope is required to sight it, brightens up very much, within a few hours or a day or two at the most. This is, somewhat loosely, called a 'nova' or 'new star'. While many of these never become visible to the naked eye, occasionally one does so and may even reach the first magnitude, or brighter, thus completely changing the aspect of the constellation in which it appears. If conspicuous, a nova is generally mentioned in the newspapers. Should the marine observer hear of one, or discover one (in which case he will usually find he is not the first discoverer) he may be interested in following its changes of brightness. The normal history of a nova is that it remains at full brightness for a short time, probably a day or two at the most, and then very gradually decreases, the reduction in brightness being interrupted by slight temporary increases. If the star has attained the third magnitude or more it may remain visible to the naked eye for several weeks.

If the observer wishes to record the exact brightness of a nova (or other star) at any time, he may select a star of about the same altitude judged to be exactly of the same brightness. If no such star is available, he should select two stars of about the same altitude as the nova, one a little brighter and one a little fainter than the nova. He can then express the brightness of the nova in terms of the small interval of brightness between the two comparison stars. For example, it might be halfway between them in brightness, or one-third of the interval, counting from the brighter to the fainter, or one-quarter of the interval, counting from the fainter to the brighter. If such an observation is received, it can be easily converted into actual magnitudes, since the magnitudes of all naked-eye stars have been accurately determined. Both these methods break down if the star is much above the first magnitude, as suitable comparison stars would probably not be available. One or more of the bright planets, if visible, might, however, serve for this purpose.

An accurate observation of the magnitude of a nova, especially in its early stages when the brightness is changing quickly, may be of great value to astronomers, since no other observation might have been made anywhere else at the same time.

SUNSPOTS

It is very dangerous to the sight to look at the sun, either with or without optical aid, without using smoked or deeply-tinted glass to reduce the light. This applies even when the sun is in partial eclipse. The only exception is when the sunlight is greatly weakened by passage through fairly thick fog, especially when the sun is at low altitude.

The number and size of sunspots varies in different years. Over a period of years solar activity, of which the occurrence of large sunspots is one manifestation, rises to a maximum and subsequently falls to a minimum. The time between successive maxima varies considerably, but averages about 11 years. For several years around the time of maximum activity, spots are frequently large enough to be seen without optical aid; sometimes two or more are thus visible at the same time. Around the time of minimum activity, spots are either very small or completely absent. The life of an individual spot may be anything from a few days to several weeks.

Owing to the sun's rotation on its axis, a spot previously formed, and coming into view at the sun's eastern limb, will appear to cross the disc in about 14 days, if it lasts so long. Apparent changes of position of the spots on the sun's disc take place during the day, but are merely due to the observer's changing angle of view. The imaginary line forming the horizontal diameter of the sun at noon appears to be tilted upward between sunrise and noon and downward between noon and sunset, the most extreme tilting occurring at sunrise and sunset.

Daily photographs of the sun through telescopes are taken at one or other of the astronomical observatories throughout the world. While marine observers may find it interesting to see the spots and note their changes of form and position on successive days, especially in years of maximum solar activity, it is not necessary to make sketches of them in the logbook as these can never be accurate enough to have any scientific value.

Solar flares. Near certain sunspots there occur areas which undergo sudden increases in brightness; these are called flares. They are best seen by means of special instruments which give a picture of the sun's surface in red hydrogen light. Some of the greatest solar flares have, however, been observed as increases in the total white light of the sun; seen in this way, a flare lasts for a few minutes and has about the same area as a large sunspot. The first such observation of a bright patch on the sun's surface was made in 1859 and several flares have been similarly observed since then. The appearance of flares cannot be predicted, but they are more numerous at times of maximum solar activity (as measured by the numbers of sunspots).

The increase in light intensity during a flare is particularly strong in the ultra-violet part of the spectrum (the part beyond the visible violet light to which our eyes are not sensitive). The blast of ultra-violet light emitted from a solar flare produces several detectable effects in the high atmosphere of the earth.

Associated with the increase in light intensity during a flare, there is ejection of material particles from the region of the flare out into space. This material shoots out at speeds of about 300 to 650 kilometres per second, which probably increases as the material gets further from the sun. If moving in the appropriate direction, this material causes interesting effects in the high atmosphere. Some of the high atmospheric effects of solar flares are described in the next chapter.

CHAPTER 10

Phenomena of the High Atmosphere

Regions of the high atmosphere. The high atmosphere of the earth is classified into regions according to the degree of ionization of the atoms of the atmospheric gases at the level concerned, that is, according to the electrical conductivity. In the lower atmosphere, air is a very good electrical insulator, i.e. its conductivity is very low. In the high atmosphere, however, many of the atoms are ionized so that the air is electrically conducting. The classification into regions depends on the different degrees of conductivity and other electrical properties at the different levels. There are three regions at the following approximate heights: the D region, between 48 and 80 km up; the E region, between 80 and 130 km; and the F region above 130 km.

High-frequency radio fadeouts. Long-distance radio transmission in the short-wave band (in the frequency range say from 3 to 30 megahertz) is normally achieved by reflection of the radiation from the F region, the main influence of the D and E regions being to absorb some of the energy of the radio waves. When the blast of ultra-violet light from a solar flare reaches the earth it increases the ionization in the D region, however, and this increased ionization causes absorption of HF signals, reducing the signal strength considerably or even causing a complete fadeout of the received signals. Such a HF radio fadeout can occur only on the daylight hemisphere of the earth; it usually lasts for about twenty minutes.

Sudden enhancements of atmospherics. Lightning flashes in the lower atmosphere are a source of radio noise, detected over a wide range of frequencies as the crackles called atmospherics. Thunderstorms and lightning flashes are most numerous in the tropics, and transmission round the earth to higher latitudes of the resultant radio noise is by means of reflection from the various conducting regions of the high atmosphere. For the low-frequency (LF) part of atmospherics (in the frequency range below 100 kilohertz) it is the D region which acts as the reflector. Following a solar flare and the consequent increased ionization in the D region, the reflection of LF atmospherics is suddenly improved and the noise level of atmospherics may rise to as much as double its normal value. The rise takes place in a few minutes and the level remains high for an hour or two.

High-frequency radio blackouts. When the material ejected from the sun during solar flares, and at other times, reaches the earth, it affects the electrical properties of the F region in such a way that it ceases to act as a good reflector for HF radio. This effect is called an ionospheric storm. It is most severe in high latitudes because the material particles coming from the sun are electrically charged and are guided to the polar regions by the earth's magnetic field. Being often, though not always, connected with solar flares, ionospheric storms are more frequent during solar activity maxima. The HF radio blackout associated with a severe ionospheric storm may last for several days.

Magnetic disturbances. The earth's magnetic field is almost entirely of internal origin, being most probably produced by electric currents flowing in the molten material of the core, but there is a small part of it (less than one-hundredth) which can be attributed to electric currents flowing in the high atmosphere. It is believed that these currents flow mainly in the E region. The daily heating of the atmosphere by the sun causes the currents to go through a regular daily cycle of change. These regular changes are very small in most places and cannot be detected by an ordinary compass needle, the maximum change of direction of the needle during a day being about one-fifth of a degree of arc.

Quicker changes occur at the times of HF radio fadeouts, but they also are too small to be detected by a compass needle. They are known as magnetic crochets or as solar flare effects.

Much larger deviations can occur, however. They are the result of large, irregular changes in the E layer currents, called magnetic storms, which occur in conjunction with ionospheric storms and HF radio blackouts in high latitudes. Like the ionospheric storms of the F region, these E region magnetic storms are caused by the arrival of the material particles ejected from the sun during solar flares and at other times.

If the magnetic storm be severe, the compass needle may be deflected continuously in one direction, to the extent of about half a degree, for some hours. In more intense storms the needle may oscillate one degree or more on either side of its normal position, and such oscillation may continue for as long as ten or twenty minutes before dying out. Further oscillation may occur after a period of quiescence. Deviations of 2° or more have been known, but are rare. During the great magnetic storm of 25 January 1938 a deviation of 4° eastward was observed off the Portuguese coast.

NOCTILUCENT CLOUDS

About one hour after sunset on clear summer nights, rather lovely clouds with delicate filigree patterns very like cirrus may be seen to appear slowly in the darkening northern sky. These are the noctilucent or luminous night clouds. What is most remarkable about the clouds is their height. They are situated at an altitude of about 80 km, far above all ordinary clouds. At this great height, the clouds remain sunlit long after sunset; they are seen by the sunlight they reflect and become visible when the sky background is sufficiently dark. They cannot be seen by day, since the light they reflect is minute compared with the light scattered by the daylight sky.

The clouds are of a pearly-white colour and often show a bluish tinge. They are classified in the following forms:

TYPE I. Veils. A tenuous, usually structureless background to other forms, rather like cirrostratus clouds.

TYPE II. Bands. Long streaks with diffuse edges (Type IIa) or sharply defined edges (Type IIb) often occurring in groups arranged roughly parallel to each other.

TYPE III. Billows. Closely spaced short bands in the form of waves.

TYPE IV. Whirls. Contorted forms indicating turbulence at the cloud level.

The clouds have been observed only in latitudes higher than 45°N and most frequently in latitudes 55° to 60°N. They appear only in the summer months and most frequently during the weeks following the summer solstice. There are indications that the distribution and behaviour of the clouds in the southern hemisphere is similar to that in the north, though, for obvious reasons, observations are few.

The clouds were at first thought to consist of volcanic dust that had been projected to great heights at the time of a great eruption and had collected below the inversion of temperature that exists above a height of 80 km. It was later claimed that volcanic dust could hardly penetrate so high in the atmosphere and that the clouds therefore were more likely to be formed from dust arising from the burning up of meteors in their passage through the atmosphere near the level of the clouds. During recent summers, however, direct evidence indicating that the cloud particles are ice crystals has been obtained from rockets carrying sampling surfaces, which were exposed as the rocket penetrated the clouds. The ice crystals are likely to have formed on a nucleus of cosmic origin. Physically therefore the clouds are no different from cirrus clouds and their appearance certainly supports this view.

Systematic observation of the clouds over the globe is only now being organized with a view to investigating more fully their distribution with respect to latitude and longitude, their spatial extent and the seasonal frequency of their occurrence.

Observers should record (a) the night of occurrence specified by two dates, e.g. 20–21 June 1968, and the latitude and longitude of the place of observation; (b) the period(s) of time, GMT, during which the clouds were observed; (c) the forms present, types I to IV as defined above; (d) the horizontal and vertical extent expressed in degrees of azimuth and elevation, at specified times, say every quarter hour, half hour, or hour during the night. This information is best conveyed by drawing a rough sketch showing the configuration of the cloud elements and the co-ordinates, elevation and azimuth of the visible boundaries of the cloud, i.e. the maximum and minimum elevations in different azimuths and the limiting azimuths, east and west of north.

Photographs are of great value. With fast monochrome film, exposure times are of the order of 15 to 20 seconds at f/3·5; with colour film of rating ASA 25, exposure times are 30 to 60 seconds at f/3·5. It is advisable to take several photographs at different exposures. The time at which photographs are taken should be recorded to the nearest minute.

Mother-of-pearl cloud. In some winters, in very clear sky, after the passage of a large, deep depression to the region of northern Scandinavia, a high form of cloud, known as mother-of-pearl cloud, has been seen in Norway, Scotland and elsewhere in north-west Europe, almost wholly within the period December to February. This cloud is of very delicate structure, somewhat lenticular in form. Its distinguishing feature is that it shows iridescence, which remains visible after sunset; iridescence on ordinary clouds never persists after sunset. The colouring is exceptionally brilliant if the angular distance of the cloud from the sun is less than 40°. They are most spectacular just after sunset, or just before sunrise. They remain visible for half an hour or more after sunset, suddenly fading when the sun sinks too low to illuminate them. Before sunrise there is a correspondingly sudden appearance.

Mother-of-pearl cloud is composed of very minute water drops. Its height is from about 19 to 29 km so that it is much higher in the atmosphere than the familiar types of ordinary high cloud. It is a rare phenomenon and its geographical distribution is very restricted. If seen, full details should be recorded including the size and shape of the clouds, their positions and angular distances from the sun, the distribution of the colours and the time at which the clouds darken in the evening, or become illuminated in the morning.

AURORAL DISPLAYS

General remarks. Associated with a severe magnetic storm there is always a great auroral display. When observing conditions are good this is one of the most beautiful and impressive of natural phenomena. As no instruments are necessary, and since reports of auroral displays are of particularly great scientific value, a fairly full description of aurora is given here, along with suggestions about methods of reporting observations.

Regions of occurrence. Aurora is primarily a polar phenomenon. It occurs most frequently in the two so-called auroral zones: these are rings of about 20° (1200 n. mile) radius, centred on the geomagnetic axis poles (which are different from the magnetic poles usually marked on maps). The northern auroral zone is centred at about 79°N, 70°W, in north-west Greenland; it runs from Cape Farewell across Iceland, over the Norwegian Sea, passing 250 n. mile north of North Cape, and over the Arctic Ocean south of Franz Josef Land to meet the Alaskan coast near the mouth of the Mackenzie River. Its highest latitude is 81°N, reached in the Arctic Ocean at about 110°E. Crossing Alaska it descends to lower latitudes, traverses Hudson Bay at about 60°N, and passes over Davis Strait to Cape Farewell again. The southern auroral zone is centred at about 79°S, 110°E and runs across Antarctica from the coast of Little America at about 150°W, through the Falkland Islands Dependencies, to the coast of Enderby Land, at about 70°E, and over the Antarctic Sea, reaching its lowest latitude, 59°S, at 110°E.

Even during quiet atmospheric conditions, there are strong electric currents flowing in the E layer around these two zones. It appears very likely that the quiet auroral arcs which seem to occur every night in the auroral zones are a visible manifestation of these currents. The process by which the gases of the upper atmosphere emit light under the influence of electric currents is similar to the process of light emission from a neon advertising tube.

During disturbed ionospheric conditions, such as follow the arrival at the earth of material ejected from the sun during a solar flare, the aurora moves from its position in the auroral zones and manifests great activity. Such an auroral display is observable from a wide range of latitudes during the course of the night. In particular, during a great ionospheric and magnetic storm, aurora is seen from subtropical and tropical latitudes, sometimes causing panic among the inhabitants of the countries where such displays are very rare. The duration of a display at low latitudes is very variable, from a few minutes to some hours, depending on the characteristics of the ionospheric disturbance concerned.

Height. It has been found that the commonest auroral forms, quiet arcs, which have a fairly distinct lower border, are nearly always situated in the E layer. The average height of the lower edge is about 100 km. An auroral arc is often several hundred kilometres in length and the height above the earth of the lower edge is the same all along it; the effect of the curvature of the earth can therefore be clearly seen, producing the characteristic arc shape.

Another very characteristic feature of auroral forms is the appearance of a folded curtain, produced by bundles of upward-stretching rays. The line of an auroral ray is found to be along the direction of the magnetic force at the place concerned. In the latitudes where aurora is most common, the lines of magnetic

87

force stand at about 20° from the vertical, inclined away from the geomagnetic pole. Rays may stretch up to heights of over 960 km, particularly just after sunset or just before dawn, when the upper parts are still sunlit, being outside the earth's shadow.

Aurora is entirely a phenomenon of the high atmosphere, where the air density is low. There are no authenticated measurements of heights less than 64 kilometres.

Variation during a night. During a great magnetic storm the time of maximum disturbance can occur at any time. Although maximum auroral activity is closely connected with maximum magnetic disturbance there is a tendency for auroral activity to reach a peak at any place within an hour or two of local midnight.

Seasonal variation. During a year there are two maxima of auroral activity around the equinoxes, which are fairly well marked. Of the twelve greatest auroral displays occurring during 1874 to 1954, three occurred in March and four in September.

Sunspot cycle variations. The great auroral displays are associated with great solar activity and therefore tend to occur around the times of maximum sunspot number. The observations made during and after the International Geophysical Year showed that in sub-auroral latitudes maximum auroral frequency coincided with the period of maximum activity of the sun (1957) and that a secondary maximum, almost as great as the primary, followed in 1959. At times of minimum sunspot number, auroral displays show the same tendency as magnetic storms to recur at 27-day intervals, these being the periods of time required for successive appearances opposite the earth of the same point on the sun's surface.

Observation of aurora. It is desirable to record all occurrences of aurora, and to give as good a description as possible, since there is much yet to be learnt about this phenomenon. Reports from ships are used in conjunction with reports from observers on land and in aircraft to compile full descriptions of all displays. These are required in connection with the study of many problems connected with long distance radio communication and other practical matters.

While auroral reports are required from all latitudes, marine observers who are familiar with auroral features can make a unique contribution to auroral studies by keeping a watch for aurora in tropical regions. Several of the great tropical displays of the past have been very poorly recorded because nearly all observers in the tropics are unaccustomed to seeing aurora and either do not recognize it at all or do not report its appearance properly. In addition, reports from all southern latitudes are very valuable, because nearly the whole of the inhabited southern hemisphere is at low latitudes where aurora is not often seen and there is therefore little auroral information available, except for that obtained by Antarctic expeditions.

The information required is as follows:

TIME. The time of each observation should be given in GMT. The times of outstanding events should be recorded to the nearest minute. Examples of such events are: the change from quiet to active forms, the onset of flaming activity, the increase in elevation of a previously stationary form etc.

ACTIVITY. When an auroral form exhibits no movement or brightness variations it is said to be quiet. When there are small irregular movements or brightness variations it is said to be active. Two characteristic types of activity are given special names. When the light from a particular auroral form waxes and wanes fairly regularly (with a period of between 10 and 100 seconds) it is said to be pulsating. The most impressive type of activity is that known as flaming, in which waves of light appear to sweep across the forms from the horizon to the zenith.

BRIGHTNESS. Four grades of brightness are recognized and are specified as follows: weak, when the light is similar in brightness to that of the Milky Way; moderate, like cirrus cloud in full moonlight; bright, like cumulus cloud in full moonlight; and brilliant, when the auroral forms appear brighter than any moonlit cloud. When the brightness is judged to be intermediate between two of these grades it can be expressed in such a way as 'weak to moderate'.

FORMS. The common forms and the symbols used to denote them are given in the drawings of Figure 29. Aurora often appears as a GLOW (Figure 29a) on the poleward horizon, almost always in the direction of the magnetic meridian; such a glow is the upper part of some other auroral form whose lower edge is below the horizon. A common form, particularly in high latitudes, is an ARC crossing the magnetic meridian; this may be HOMOGENEOUS, i.e. uniform in brightness (Figure 29b), or RAYED, i.e. with vertical ray-structure (Figure 29c). The reason for the arc shape is given above, under Height. Multiple arcs, running parallel to one another across the sky, are not uncommon. The sky between an arc and the horizon may appear to be darker than the surrounding sky at the same altitude; this is only a contrast effect, and stars may be seen undimmed in the so-called 'dark segment'. When the form has not the regular shape of an arc but has folds along its length it is called a BAND: this also may be HOMOGENEOUS (Figure 29d) or RAYED (Figure 29e). Rayed bands in which the rays are long look like curtains or draperies. An active rayed band overhead is perhaps the most impressive of all auroral forms, particularly if there are colour changes in the waving folds. Single RAYS (Figure 29f) or bundles of rays are often seen after the break-up of a rayed arc or band. When such rays rise from behind a surface feature, a searchlight effect is produced. All rays in any display are very nearly parallel to one another, but perspective causes convergence and produces fan-like formations. Rays overhead always appear to converge to a point called the magnetic zenith, which is displaced from the overhead zenith towards the equator by an amount depending on the distance from the geomagnetic axis pole. For the British Isles, for example, the magnetic zenith is about 20° south of the overhead zenith. When rays surround this point, the rays are said to be CORONAL (Figure 29g). Particularly towards the end of a display the aurora appears in PATCHES (Figure 29h), which are like diffuse clouds, without any arc-type or ray-type structure. In lower latitudes an auroral display often consists largely of deep red patches.

ELEVATION. This is measured as altitude in degrees from the horizon. The most important measurement is that of the elevation of the lower edge of an arc at its highest point. This can be indicated by the letter h; thus $h = 25°$ means that the highest point of the lower edge of an arc is at an altitude of 25°.

DIRECTION. It is usually sufficient to give this in terms of true compass points, but accurate directions of features such as isolated bright rays are of value; these should be given in degrees of true azimuth. A sketch with angles marked on it is often easier to make and to read than a description in words.

COLOUR. Very often the intensity of auroral light is too low to stimulate the colour-sensitive part of the eye and all the forms appear pale and grey, like clouds illuminated by weak moonlight. But when the intensity increases, a variety of colours is seen. The commonest is yellow-green, one of the character-istic colours emitted by oxygen gas under the conditions prevailing in the iono-

(a) Glow

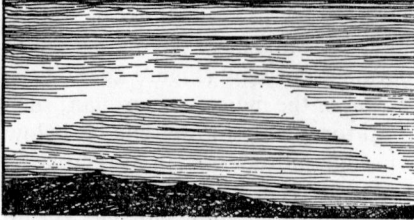

(b) Homogeneous arc

(c) Rayed arc

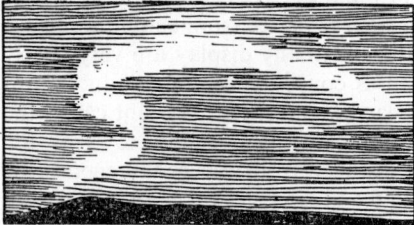

(d) Homogeneous band

(e) Rayed band

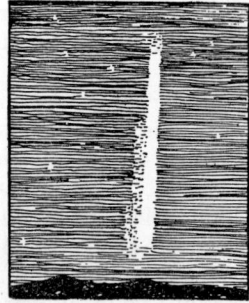

(f) Ray

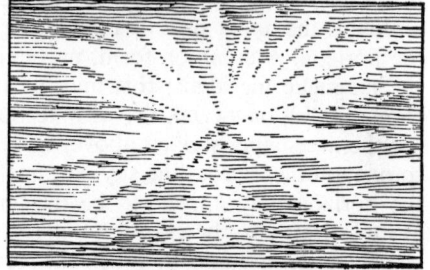

(g) Coronal rays

(h) Patch

FIGURE 29. Auroral forms

90

sphere. Sometimes a red coloration is predominant. This usually comes from oxygen also, but in aurora at very low levels red nitrogen light has been found. Nitrogen, however, gives mainly blue and violet shades; and these are strongest in the long sunlit rays which occur just after sunset and just before dawn. Many mixtures of these and other colours are possible.

Photography of aurora. Because auroral light is of low intensity, time exposures are required for photography and this is not often possible on shipboard. Given steady conditions, an exposure of 15 to 30 seconds, with a lens aperture of f/3·5 and a fast film, should give a reasonable image of a bright auroral feature. With colour films of rating ASA 25, exposure times are 20 to 60 seconds at f/3·5 depending on the brightness of the feature. It is advisable to take several photographs at different exposures.

AIRGLOW

On a clear starlight night, in the absence of normal or abnormal twilight, moonlight, lunar twilight, thin high cloud over the sky, auroral displays or artificial illumination from towns etc., the sky background is not dark, but has a certain degree of luminosity. While some of this luminosity is due to the combined light of stars too faint to be seen individually without the aid of a telescope, the greater part is due to a faint glow known as the airglow. Older names for this were 'permanent aurora' and 'earthlight'.

The airglow is generally uniform over the sky except towards the horizon where it is usually somewhat brighter. The intensity is not always the same on different nights and there are exceptional nights when the sky background appears to be unusually light. There are no means of estimating the intensity of the glow by visual observation, so that the phenomenon is not one which can be usefully observed at sea.

METEORS AND METEORITES

During the night watches the seaman has many opportunities for obtaining useful observations of meteors. 'Meteor' is a general term to include all those small bodies which, travelling through space, encounter the upper part of the earth's atmosphere and become visible by the incandescence produced by the friction of very rapid passage through it. It is estimated that many thousands of millions of these objects enter the atmosphere daily, but all except a minute fraction of these are much too faint to be naked-eye objects. The vast majority of all meteors are entirely disintegrated and subsequently settle down slowly to the earth's surface in the form of extremely fine dust. A number, however, of greater size, may more or less wholly survive the disintegrating effect and fall to the earth's surface as solid objects of varying sizes. These are called meteorites. On rather rare occasions they have been observed to fall into the sea, the actual splash being seen. Such an event should be recorded in the logbook in full detail. No meteor should be described as a meteorite unless it is definitely known to have reached the land surface or have fallen into the sea.

The ordinary small meteors appearing as luminous streaks in the sky are popularly known as 'shooting stars', but they have of course no connection with any star, being merely small fragments of matter varying in size from grains of sand to a pea. Larger objects are known as fireballs. These may be as bright as a first-magnitude star or may equal the planets Jupiter or Venus, or, in the very finest examples, may greatly exceed the full moon in brilliance. Objects which fall as meteorites usually appear first as one of these very bright meteors, but the reverse does not hold good, as the majority of the brighter meteors are completely disintegrated in the air.

Appearance and speed of meteors. The appearance of meteors is as varied as their brightness. Some travel fast, others slowly. The apparent path is usually a straight line or arc, but it may assume other forms. Some leave streaks of sparks or luminous vapour, known as trails. In many cases the trail disappears immediately, but in others it remains visible for seconds or minutes, or in rare cases for periods up to two hours or more. When the trail remains visible for some time, changes may be observed in it, caused by the air currents of the high upper air, combined with the fall of the material due to gravity. Most large meteors and fireballs are strongly coloured and the colours may change during the flight. Sometimes the meteor appears to break up, the detached portions then proceeding separately, or it may appear to explode at the end of its course. The trails are usually reddish, white or golden, whatever the colour (or changing colours) of the meteor itself.

The duration of a meteor's flight is rarely more than three seconds, and is apt to be greatly overestimated. The actual speed of a meteor when it enters the earth's atmosphere is usually between 18 and 84 kilometres per second. The average height above the earth's surface is 120–130 km at the time of appearance and 70–80 km at the time of disappearance. The height of the beginning and end of a meteor's visible path in the atmosphere and its speed are determined by observations made by two observers some distance apart, up to 100 n. mile or more. At much greater distances the same meteor could not be seen by both observers, since an individual meteor is only visible over a small part of the earth's surface and would thus be below the horizon of one of the observers. In this joint observation each observer notes the points of appearance and disappearance of the meteor in the sky as accurately as possible, and the duration of its flight. The information derived from such observations is valuable, not only in extending our knowledge of meteors, but also in making inferences about the temperature of the very attenuated atmosphere at very great heights above the earth's surface.

Frequency of meteors. Meteor showers. The number of meteors seen in a given time is usually greater on nights of higher atmospheric transparency, since more of the fainter meteors are seen, and these are much more numerous than the brighter ones. On nights of equal clarity, about twice as many are visible in July to December as in January to June. Furthermore on any single night of the year the hourly rate of meteor appearance is greater after midnight than before it. These remarks refer to average conditions. On certain nights it may be seen that meteors are more numerous than usual and that their tracks, produced backward, would all converge to the same point or small area in the sky. Such a group of meteors, with the same radiant point, constitute a meteor shower. Many of these occur annually, though not always with the same intensity; some recur only after an interval of a number of years, and others are unpredicted and unexpected. Prolific showers, with perhaps many thousand meteors per hour, such as were given by the Leonids in November of 1799, 1833 and 1866, have been extremely rare of more recent years. The collection of meteors forming a shower are moving in the same general orbit in space and in a few cases the orbit of a shower has been found to be the same as that of a known comet, of whose material the meteors originally formed a part.

Observation of meteors. The complete observation of a meteor comprises:

(a) The positions of the points of appearance and disappearance in the sky.
(b) The duration of the flight.
(c) The magnitude of the meteor relative to a named star or planet, or a general estimate, such as first magnitude.
(d) Any notable colour, colour changes, persistence of trail, or other peculiarities.

It is not necessary to record the meteorological conditions at the time of a meteor observation, but if the flight is only partially seen, owing to cloud, this should be noted.

Owing to the suddenness of a meteor's appearance, it is often difficult to fix in mind the points of appearance and disappearance. This should be done as accurately as possible with respect to neighbouring stars, and from a star atlas. The right ascension and declination of each point can then be found. Alternatively, the position can sometimes be given as an angular distance from one named star in the direction towards another named star. Such observations of position, if reasonably accurate, say to the nearest half-degree, are of use in determining the radiant point of the meteor or for combining with another observation made at sea or on land, to find its height and speed. Positions by azimuth and estimated altitude are hardly accurate enough for these purposes. An observation of the brightness and appearance of a fine meteor is, however, of interest, even if its track in the sky has not been exactly determined.

Observations of a persistent meteor trail are of interest. The shape of the trail and its position relative to at least two named stars should be drawn at suitable intervals of time until it disappears. If carefully timed observations of the same trail are received from two or more observing ships, some information about the speed and direction of the wind at the time in the high atmosphere can be computed.

The appearance of meteors in unusual numbers on any night, especially if obviously directed from the same point in the heavens, should be put on record. The occurrence or non-occurrence of showers on a particular night is often of considerable astronomical interest, and it may happen that the conditions are such that a meteor shower is visible only in restricted longitudes.

ARTIFICIAL SATELLITES, RESEARCH ROCKETS AND ALLIED PHENOMENA

Artificial satellites. In August 1975 there were approximately 3500 earth-orbiting payloads and debris of which about 300 were observable with the unaided eye by the sunlight reflected from them at times when the observer was in darkness or twilight. A satellite can normally be distinguished from a planet or meteor by its angular velocity across the sky, which is much greater than the former and much less than the latter. In addition, the length of time for which a meteor can be seen is seldom greater than 3 or 4 seconds, while a satellite may be visible for up to about 45 minutes and a planet up to about 12 hours. So long as they behave in a normal manner, i.e. steady course and speed, the passages of satellites are not of great importance and need not be recorded in the meteoro-logical logbook. A special case, however, is the re-entry of a satellite into the earth's atmosphere and its burning up. The accurate observation of this event may yield information about the density of the atmosphere through which the satellite is travelling.

Re-entry occurs when a satellite, due for example to the cumulative effect of small drag forces throughout its lifetime, descends into the denser regions of the high atmosphere. The frictional effects increase as a consequence, it descends still further and its path about the earth rapidly degenerates into a spiral. During the final stages of this process the satellite becomes incandescent because of its extremely high temperature and eventually it disintegrates and burns up.

In appearance, a re-entering satellite is similar to the passage of a meteor but it is brighter, moves more slowly across the sky and may be visible for several minutes. A re-entry can happen in any part of the world and, when seen, should be recorded by following the general rule outlined in Chapter 8. When recording a possible sighting of a re-entering satellite it is important to note as accurately as possible the direction of travel of the object across the sky, its brightness compared with known stars, the time, bearing and elevation angle when first and last seen, together with any extraneous circumstances such as intervening cloud which may have curtailed or otherwise limited the observations.

Research Rockets. These are launched from various ranges such as Thumba in 8° 32′N, 76° 52′E and carry equipment for research into the environment of the earth, including its high atmosphere, or for astronomical observations. They are usually projected vertically or nearly so and, after attaining their maximum height which may be 160 km or more, fall back to the earth's surface. The direct light from the burning propellent during the upward leg of the flight will be readily identified but some of the experiments carried by the rockets may themselves give rise to effects which are visible at night. For example, equip-ment may be carried which releases chemicals into the high atmosphere.

Some chemical vapours are ejected in the form of a long bright trail on the upleg or downleg of the rocket path. Night-time trails usually appear white but at twilight, about 30 minutes after sunset or before sunrise, the trails are generally yellow, red or greenish-blue. Although their size depends on the altitude at which they are released and, of course, on the position of the observer, it is not unusual for a trail to be as much as 40° long and from 1° to 5° wide. These vapour trails, which may remain visible for more than 20 minutes, are generally distorted by winds in the upper atmosphere.

Sometimes 'grenades' are ejected from the rocket during the upward flight, generating acoustic waves whose reception at the ground enables the wind and temperature structure of the upper atmosphere to be calculated. At night a grenade burst would appear similar to an AA shell burst, being much brighter than a planet but of very short duration. After 2 or 3 seconds another burst would appear displaced in the direction of flight. The number of bursts may total about 20. No sounds are likely to be heard as the acoustic waves are generated at sub-audible frequencies and are very weak in intensity on reaching the ground. In some rocket firings the grenade ejections are extended above 100 km altitude and the gas products of the explosion then react chemically with the atmosphere, producing a faintly luminous spherical cloud about the size of the full moon although very much fainter. Such clouds may last for 5–10 minutes before diffusing and disappearing and during this time the wind speed and direction can be found from their drift. On other occasions a faintly luminous trail may be produced for the same purpose by the release of a certain chemical from the rocket. The trail, like the grenade clouds, is a soft white in colour. Such experiments are often held under twilight conditions with sunlight falling on the releases against a still dark sky. The sunlight is re-radiated in certain parts of the spectrum from the cloud or trail providing a spectacular blue-green hue, or the well-known yellow coloration of sodium light, if sodium vapour has been released.

CHAPTER 11

Phenomena of the Lower Atmosphere

The presence of water droplets, ice crystals and dust particles in the atmosphere and local abnormalities in the distribution of temperature and humidity give rise to a wide variety of interesting and beautiful phenomena which are mainly due to either the reflection, refraction or scattering of the rays of the sun or moon. Many of these 'optical phenomena' are described below. Also included are descriptions of a few phenomena which are due to other causes and which are equally worthy of careful observation when they occur.

ABNORMAL REFRACTION AND MIRAGE

A mirage is produced by refraction of light in the layers of air close to the earth's surface. Two main classes of mirage occur, (a) inferior and (b) superior, in which the virtual image is below and above the object, respectively. The inferior image is seen over a flat, strongly heated surface (e.g. desert) and gives the illusion of an expanse of water; it is caused by the strong upward reflection of light from the clear sky towards the observer. The superior mirage is seen above a flat surface of much lower temperature than the air above it; light from an object is, in this case, bent downwards towards the observer, as in 'looming'. In such physical conditions multiple reflections may give rise to various images, some displaced laterally with respect to the object.

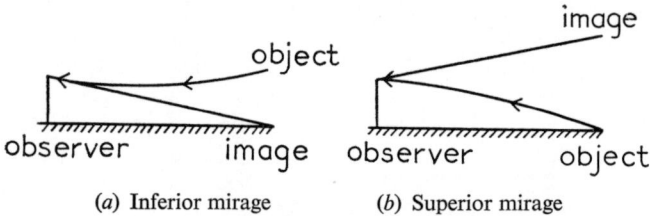

(a) Inferior mirage   (b) Superior mirage

FIGURE 30. Mirages

Good descriptions and sketches of the various forms of mirage and the effects of abnormal refraction are always of interest, especially of the more striking forms, such as a well-developed superior mirage. Unusual phenomena should be carefully reported, such as the apparent discontinuity or distortion of the horizon line that has been occasionally seen, also lateral mirages and the complicated mixed mirages of the Strait of Messina, known by the name, of Italian origin, of 'fata Morgana'. When lights are seen at abnormal distances, the normal distance of visibility should be given. In all observations or abnormal refraction and mirage, the temperature of air and sea, the type and amount of cloud present, and the direction and force of the wind should be noted.

GLORY OR BROCKEN SPECTRE

In a foggy atmosphere an observer, standing with his back to the sun, when this is at low altitude, will sometimes see the shadow of himself, or of his head, thrown upon the fog, together with coloured rings of light surrounding the shadow. The phenomenon was first noted on the Brocken mountain in Germany but it is not confined to mountain districts and it is most common in Arctic regions, where it is seen on every occasion of simultaneous sunshine and fog.

The coloured rings are now usually known as a 'glory'. A typical series of colours seen in a well-developed one is as follows. There is a general whitish-yellow colour round the shadow, surrounded by rings of colour in order outwards: dull red, bluish-green, reddish-violet, blue, green, red, green, red. A white rainbow at a considerable distance outside the glory is sometimes also seen.

The shadow of the observer on thick fog may be seen at night if there is a bright artificial light behind him.

COLOURED SUNS AND MOONS

The various red or orange colours ordinarily exhibited by the sun or moon when near the horizon are generally caused by these bodies being viewed through a great thickness of the dust-laden lower atmosphere, which absorbs most of the sunlight of shorter wave-lengths, leaving the longer ones, mainly yellow and red, to come through.

Occasionally in twilight the moon appears to be of a greenish colour, usually a pale greenish-blue or a pale apple-green colour. This is an effect of colour contrast when the twilight hues of the surrounding sky are brighter than usual, either purplish or reddish, or when the moon is near or covered by thin, brightly-tinted cloud.

Coloured suns or moons, not an effect of colour contrast, are sometimes seen. This phenomenon may be produced by dust or smoke haze in the lower atmosphere, e.g. a scirocco laden with dust from the Sahara may give a blue sun or moon in the Mediterranean, and a similar colour may be given in the region of extensive bush fires. The phenomenon may also be produced by volcanic dust at high atmospheric levels. Blue and green moons were observed on many occasions after the great eruption of Krakatoa in 1883 and the sun assumed many different and often quite brilliant colours. Shades of red and copper, green, golden-green, blue, both silvery and leaden, were seen on various days in different localities.

Coloured suns and moons were seen over much of western Europe between 26 and 30 September 1950. These were produced by the smoke from an extensive forest fire in Alberta, Canada which began on 23 September. The sun's colour was observed in different places as steel-grey, deep blue and purple.

Any observations of this kind are of interest.

CORONAE

A corona consists of one or more coloured rings round the sun or moon as centre, when this is covered with middle or lower cloud of sufficient thinness to allow the greater part of the light to come through. It is distinguished from a halo by its smaller size and different colouring, as explained below. A fully developed corona shows a bluish-white or yellowish glow, usually 2° or 3° in diameter, round the sun (or moon). Outside this is a brownish-red ring. The inner glow and the brownish ring together constitute what is called the aureole. Outside this are coloured rings, in the opposite colour sequence to that of a halo,

97

viz., violet or blue nearest the sun and red farthest out. Sometimes the whole of this colour sequence is repeated outwards a second or, on rare occasions, even a third or fourth time. A corona showing the outer coloured rings is comparatively infrequent, but the aureole alone is the commonest of optical meteorological phenomena and is formed, at any rate partially, whenever broken cloud edges of cumulus, stratus or stratocumulus pass over the sun or moon.

While the radii of the various haloes are constant, that of a corona varies on different occasions, being dependent on the size of the water-drops in the cloud. The outside radius of a fully developed corona is usually much smaller than that of the 22° halo, and is generally between 5° and 8°. After great volcanic eruptions, when fine dust is suspended at great heights in the atmosphere, an aureole comparable in size with the 22° halo has been seen; it is known as Bishop's ring.

Faint coronae are visible round the bright planets, Venus and Jupiter, and also Mars when this is sufficiently bright, providing the cloud is very thin. They may sometimes be seen round the brightest stars, especially if binoculars are used.

A yellowish blur 2° or 3° in diameter is often seen round the sun or moon and is sometimes formed by higher cloud than that which normally gives coronae. Although it has a fairly sharply defined circular edge it must not be regarded as an aureole unless bounded by the characteristic brownish-red ring.

In certain circumstances the sun or moon may show a halo and a corona simultaneously.

The name 'corona' is also given to the outer part of the sun's atmosphere (see page 78); this is directly visible only during a solar eclipse and is distinguished by the term 'solar corona'.

CORPOSANT

The electrical phenomenon known as Corposant or St Elmo's Fire is not infrequently observed at sea during squalls and thunderstorms. It is a luminous apparition seen at the extremities of masts and sometimes on the stays, aerial, jackstaff or other parts of the ship. It may appear as a brush discharge of radiating streamers several inches long, or as luminous globes, a number of which are sometimes seen along the aerial. At other times a structureless glow envelops an elongated object, such as a mast or an aerial. St Elmo's Fire is usually bluish or greenish in colour, but a violet glow has been reported and sometimes the colour is pure white.

CREPUSCULAR RAYS

The word 'crepuscular' means 'associated with twilight'. Occasionally, soon after sunset, the clear sky appears to be divided into lighter and darker rays by lines diverging from the position of the sun below the horizon. The lighter rays are those illuminated by sunshine; they are usually coloured pink, but may, on different occasions, show some shade of red or orange. The darker rays are shadows, from which the sunlight is cut off by clouds near or just below the horizon or by the irregularities of hills and mountains on the horizon. They appear greenish by contrast with the pink rays.

As the light rays come from the sun, and so are practically parallel, their apparent divergence is an effect of perspective. In favourable circumstances the light rays and shadows extend right across the sky and appear to converge, by perspective, to a point a little above the eastern horizon. These 'anti-crepuscular rays' are generally ill defined.

It is not necessary to record this phenomenon in the logbook unless it shows some feature of special interest, such as unusually distinct colouring or a well-defined convergence to the eastern horizon.

On rare occasions one or more bands have been seen extending up into the sky, from the western horizon, at a later stage of twilight. They appear of a deep blue colour, darker than the general blue of the sky, and are probably shadows of mountains well below the horizon. It is of interest to record these observations.

There are two other allied phenomena which are frequently seen and are of no special interest, unless some unusual feature is observed. The first consists of pale blue or whitish rays diverging from the sun in the day-time when it is behind cumulus or cumulonimbus cloud. The rays are sharply defined and separated by deep blue bands, which are the shadows of parts of the irregular cloud-edge. The second is associated with stratus or other cloud obscuring the sun. If there are small gaps in the cloud, sunbeams pierce these, directed more or less downward, and are rendered luminous by mist or dust in the air. This is popularly known as 'the sun drawing water'.

DUSTFALL AT SEA

Dust from the land may be blown over the adjacent sea by high winds, but not normally in appreciable quantity. In special regions, e.g. the Red Sea, sand or dust storms are not infrequent and are sometimes severe.

Desert dust or sand may be carried up to high levels of the atmosphere and finally be dispersed over so great an area as not to form any perceptible deposit on falling. The desert dust from Australia carried north-westward by the south-east monsoon reduces visibility over the East Indies region but is not observed as dustfall.

On the other hand, falls of fine reddish or brownish dust from the Sahara, carried by the trade wind, are experienced over a large area of the eastern North Atlantic adjacent to the coast of Africa, centred roughly on Cape Verde Islands. At times this deposit may lie quite thickly on board ship. Visibility in this area is often poor; not infrequently the sun appears blood-red and at night all but the brightest stars at high altitudes are obscured.

Considerable or heavy dustfall may be experienced after a great volcanic eruption. Dust from the eruption of Krakatoa in 1883 was collected on shipboard in the Indian Ocean at a distance of 1000 nautical miles. After the eruption of Hekla in March 1947, dust was similarly collected at a distance of 450 nautical miles. In August 1966 the dust from the eruption of Mount Awu in the Sangihe Islands covered the decks of a ship 225 nautical miles away in the Celebes Sea.

THE GREEN FLASH

At sunset, the small segment of the upper part of the sun's disc, which is the last to disappear, may turn emerald-green or bluish-green at the instant of its setting. The phenomenon thus usually lasts only a fraction of a second, which is the reason for calling it the 'green flash', but longer durations of the colour are occasionally seen.

The green flash is not always seen and when it is seen it is not always equally brilliant. It can range from a green of extreme brilliance and purity, conspicuous without optical aid, down to a trace of grey-green coloration observable only with binoculars.

The green flash is produced by the last rays of sunlight emanating from the upper limb of the sun, at sunset, being diffracted before reaching the observer's eye. The shorter waves which appear as violet, blue and green light suffer greater refraction than the orange and red longer waves of the white sunlight. The fringes of the upper limb cannot usually be seen while the main body of the sun is still above the horizon, as the general sunlight is too strong, but when most of this is cut off by the horizon they spring suddenly into view. Normally, only the green fringe is seen, the light of still shorter wave-lengths usually being scattered by its horizontal passage through the lower atmosphere. The flash is, however, occasionally seen as a blue one, or as green quickly changing to blue. On very rare occasions the violet colour has been seen.

The green colour occasionally appears in other ways. Sometimes when refraction is marked, and the sun's disc is perhaps distorted, the use of shaded binoculars will show that the upper limb appears to be 'boiling', giving off shreds or tongues of green 'vapour'. Occasionally the sun's upper limb has been seen with a narrow green rim when half or more of the disc remained above the horizon.

A sea horizon is not essential for observing the green flash; it may be equally well seen when the sun sets behind a distant land surface. It has also often been seen when the upper limb sinks below a bank or bar of hard-edged cloud at low altitude, and if there are several parallel bars of cloud in clear sky the phenomenon may be seen more than once on the same evening. When the lower line of the sun appears from behind cloud near the horizon the converse phenomenon, the 'red flash', has sometimes been seen.

The green flash occurs also with the moon, but has seldom been observed, presumably because it is fainter and rarely looked for. On the other hand, it has been frequently seen at the setting of the bright planets Venus and Jupiter, and an observation of a blue flash from Venus is on record. Many interesting varieties of phenomena may occur before these planets set, the observation usually requiring binoculars. Colour changes may be seen, usually between white, red and green, or two images may appear of the same or different colours. The planet may exhibit slow 'shimmering' movements, obviously due to abnormal refraction.

The most favourable conditions for seeing the green flash, at any rate brilliantly, is probably some degree of abnormal refraction, whereby the vertical extent of the colour separation described above is greater than that produced by normal refraction. In addition, the green flash is most likely to be seen when the air is relatively dust-free, and without mist or haze, so that the sun remains brighter and less red than usual at low altitudes. The green flash has been well observed at sunrise, but less frequently, perhaps because it is less often looked for. Also, owing to its short duration the phenomenon is liable to be missed unless the exact spot at which the sun will appear is known.

The green flash has sometimes been called, rather inappropriately, the 'green ray'. It will be obvious from the remarks made above that it exhibits a considerable variety of appearances at different times. Further observations, giving as much detail as possible, will be very useful in increasing our knowledge

100

of this interesting phenomenon and the conditions most favourable for its appearance.

Other phenomena involving green coloration of the sky in the vicinity of the sun at the moment of sunset are occasionally seen, and observations of these are also of interest, as they exhibit much variety. Some examples are (a) a momentary ray of green light shooting up into the sky, sometimes to a considerable altitude at or just before the final instant of sunset, (b) an appearance resembling a rapidly rotating green searchlight beam, (c) a transitory appearance as of green mist in the sky above the setting sun.

HALO PHENOMENA

Halo phenomena: *a group of optical phenomena in the form of rings, arcs, pillars or bright spots, produced by refraction or reflection of light by ice crystals suspended in the atmosphere (cirriform clouds, ice fog etc.).*

These phenomena may show colours when formed by refraction of the light from the sun, but halo phenomena produced by the light of the moon are always white. The many different kinds of halo which have been observed may be described by reference to Figure 31. This is a composite diagram made up from a number of drawings of an unusually complete halo display seen at about midday on 6 March 1941 in various localities in the west Midlands.

The four rings described below (three haloes and the parhelic circle) may under very favourable conditions appear complete but they are more frequently incomplete. Parts of these rings, together with the arcs and mock suns described on pages 102–103, may at times be seen with no apparent connection with one another; the sky may then assume a very strange appearance.

Halo of 22° (small halo) is the most frequent halo phenomenon and appears as a luminous ring, F, in the figure, with the sun or moon, S, as centre, and having a radius of 22°. The space within the ring appears less bright than that just outside. The ring, if faint, is white; when more strongly developed it shows coloration; the edge nearest the sun is red and this is followed by yellow and, in some rare cases, a green or violet fringe can be detected on the outside.

The angle of 22° is the angle of minimum deviation for light passing through a prism of ice with faces inclined at 60°, and this halo is probably due to the refraction of light through hexagonal prisms among ice crystals in cloud.

Arcs of contact to 22° halo. Among the phenomena which, from their manner of formation, can be seen only as arcs, are the so-called arcs of contact. Two of these are shown in Figure 31—an arc of upper contact, J, and an arc of lower contact, K. The arc of lower contact is very rare and so also are contact arcs which occasionally appear at the sides of the halo.

When the sun is low the arcs of upper contact appear with their convex sides turned towards the sun. The points of contact may appear as mock suns, though not belonging to the mock-sun ring proper, and they are very luminous. They may display brilliant colour effects, with the red turned towards the sun and thus being on the convex side of the arc.

When the sun is high the arcs of upper and lower contact may appear concave to the sun. Very rarely the ends are joined to form a circumscribing halo which is approximately elliptical.

Halo of 46° (large halo). The halo of 46° (large halo, G in Figure 31) is occasionally seen, though it is seldom complete. It is much less common than

101

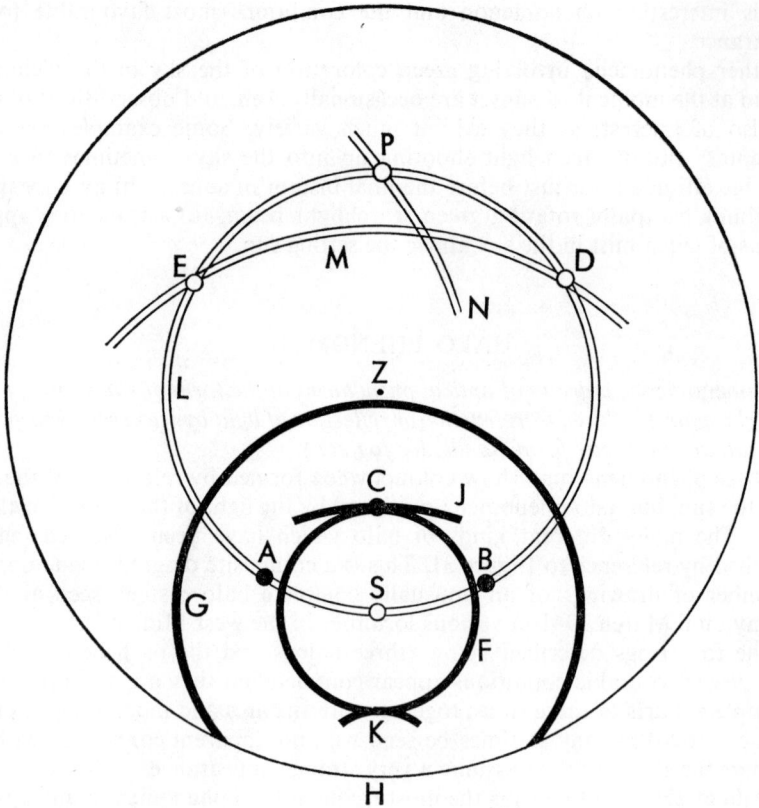

FIGURE 31. Halo phenomena

An unusually complete halo display seen at about midday on 6 March 1941 in the west Midlands; composite diagram based on observations from various localities. Appearances due to refraction, which may be brilliantly coloured, as on this occasion, are shown black; appearances due to reflection, which are always white, are shown by the finer lines. The outer circle, H, is the complete horizon, with Z, the zenith, in the centre; S is the sun, P the anthelion, A, B, C, D, E are parhelia or mock suns; F is the 22° halo, G the 46° halo, J, K are upper and lower arcs of contact of F; L is the parhelic circle or mock-sun ring (parallel to the horizon), M the arc through the mock suns at 120° (usually a pair of arcs not joined in the middle), and N the oblique arc through P (one of a symmetrical pair which may sometimes be seen together).

the halo of 22° and is always less bright. This halo also has arcs of contact; in fact these arcs occur more frequently than the halo itself and are sometimes mistaken for it. This halo requires crystals with faces at right angles.

Parhelic circle (the mock-sun ring). Occasionally a white ring which passes through the sun parallel to the horizon may be recognized. This is called the horizontal or parhelic circle or the mock-sun ring. It is shown in Figure 31 in its complete form, L; frequently, however, the minor arc passing through the sun is not visible. This portion, cut off by the intersections with the halo of 22° may in fact be distinct, faint, or invisible on occasions when the major arc, or parts of it, can be clearly seen. Bright spots may be observed at certain points of the parhelic circle. These bright spots occur most commonly a little outside the halo of 22° (parhelia, at A and B), occasionally at an azimuthal distance of 120° from the sun (paranthelia, at E and D), and, very rarely, opposite the sun (anthelion, at P). The corresponding phenomena produced by the moon are called paraselenic circle, paraselenae, parantiselenae, and antiselene. When the parhelia, paranthelia, or the anthelion are particularly bright they are called mock suns; paraselenae, parantiselenae and the antiselene, when bright, are called mock moons.

Mock suns shown at A and B in the figure are very luminous and brilliantly coloured. Red is on the side nearest to the sun, with yellow, green, blue and violet following; the blue is generally indistinct and the violet usually too faint to be distinguished. These mock suns are situated approximately on the 22° halo when the sun's altitude is 10° or less; with increasing altitude they are formed further outside the halo, being 14° outside when the sun's altitude is 55°. They cannot be formed when the sun's altitude exceeds 60° 45′. They lie on the parhelic circle which may or may not be visible at the time. Mock suns at E and D in the figure are white. The image of the sun occasionally observed, at P in the figure, is a brilliant white and is sometimes termed the 'counter-sun'.

Through the mock suns, D and E in the figure, two separate arms, the paranthelic arcs, may rarely be seen and they are sometimes (though very rarely) observed to join and appear to form the continuous arc M. The oblique arc N, through the anthelion P, is one of a pair; on some occasions one may be clearly seen while the other is invisible. These arcs, being caused by reflection, are white.

Halo of 90°. A fourth ring, the halo of 90°, is exceedingly rare; it is not shown in the figure. The ring is white and cannot be seen in its entirety unless the sun is in the zenith.

Circumzenithal arcs. Occasionally the upper and lower circumzenithal arcs may be observed; they appear to lie in horizontal planes. The upper circumzenithal arc (brightly coloured, with red on the outside and violet on the inside) is a rather sharply curved arc of a small horizontal circle near the zenith; the lower circumzenithal arc is a flat arc of a large horizontal circle near the horizon. The upper arc occurs only when the angular altitude of the luminary is less than 32°; the lower arc occurs only when the angular altitude of the luminary is more than 58°. The upper arc touches the large halo, if visible, when the angular altitude of the luminary is about 22°; the lower arc touches the large halo when the angular altitude of the luminary is about 68°. The arcs become increasingly separated from the large halo as the angular altitude of the luminary departs from the above values. Circumzenithal arcs may be observed without the large halo being visible.

Sun pillars may also be seen occasionally, particularly at sunrise or sunset. They frequently extend about 20° above the sun and generally end in a point. At sunset they may be entirely red, but are usually a blinding white and show a marked glittering. If the sun is high in the heavens they may appear as white bands vertically above or below it, but they are not then very brilliant and are often short. Occasionally, however, these white columns appear simultaneously with a portion of the white mock-sun ring, and so form another remarkable phenomenon, the cross. Sun pillars are due to reflection of sunlight from ice crystals.

The undersun is a halo phenomenon produced by reflection of sunlight on ice crystals in clouds. It appears vertically below the sun in the form of a brilliant white spot, similar to the image of the sun on a calm water surface. It is necessary

103

to look downward to see the undersun; the phenomenon is therefore only observed from aircraft or from mountains.

The observation of haloes. High latitudes, especially the polar regions, are the most favourable for frequent and brilliant displays of halo phenomena, which can be formed not only by cirrostratus cloud, but also by ice fog. Many fine displays occur, however, in temperate latitudes, where the late spring is an especially good season.

Cirrostratus is the most favourable cloud for the production of halo phenomena; the thinner and more uniform its texture the better. On the most suitable occasions, the blue sky is only dimmed with a uniform milky appearance. When the cloud is thicker in some places than others, and especially when wisps and streaks of cirrus are mixed with it, not only are the pehnomena less distinct but straight or curved lines of cloud may be mistaken for additional halo phenomena.

When thin cirrostratus is present and one or more of the commoner halo phenomena are well seen, the prospects of seeing some of the rare halo phenomena are good and a careful general look over the whole sky may result in something else being seen. Attention should chiefly be concentrated on the following regions (a) that surrounding the sun up to a radius of at least 46°, (b) a belt of the sky, at the same altitude as the sun, all round the horizon, (c) the overhead sky, with the zenith as centre.

On account of the methods of formation of halo phenomena, the reflection and refraction of light by ice crystals, the position of each halo etc., is always precisely the same relative to the sun or, in some cases, to the zenith. A halo phenomenon is thus identified by its position in the sky; its appearance is of secondary importance, though, in some cases, this helps in the identification. The most essential part of a halo observation is therefore the determination of its position by angular measurement with reference to the sun (or moon) or, in appropriate cases, the horizon or the zenith. Most of the rarer phenomena can only thus be identified with certainty.

The altitude of the sun, to the nearest degree, should also always be given, since this affects the precise position of certain halo phenomena, and in some cases determines what phenomena it is possible to see at the time. The radius of the relatively well-defined inner edge of any halo, or part of a halo, centred on the sun should be measured in degrees from the sun's centre. In the case of arcs situated vertically above the sun such as the circumzenithal arc, the distance of the lowest part of the arc from the sun is all that is required. It is useful, however, to estimate the extent of any such arc as a fraction of the small circle of which it forms part.

The mock-sun ring is identified by its parallelism to the horizon, at the sun's altitude; no measurements are required. A phenomenon situated on it, such as the anthelion or other bright spot, or the point of intersection of an arc with it, is measured in the form of azimuth distance from the sun.

The above statement should be sufficient to indicate to the observer the lines on which he should proceed. The most difficult cases are certain abnormal phenomena such as are shown in Figures 32 and 33. The diameter of any halo not centred on the sun or moon could be measured by sextant; the altitude and azimuth of the estimated centre of the halo would then give its position. The position of any detached arc could be measured by taking the altitude and azimuth of each of the two ends, and of the point on the halo equidistant from these.

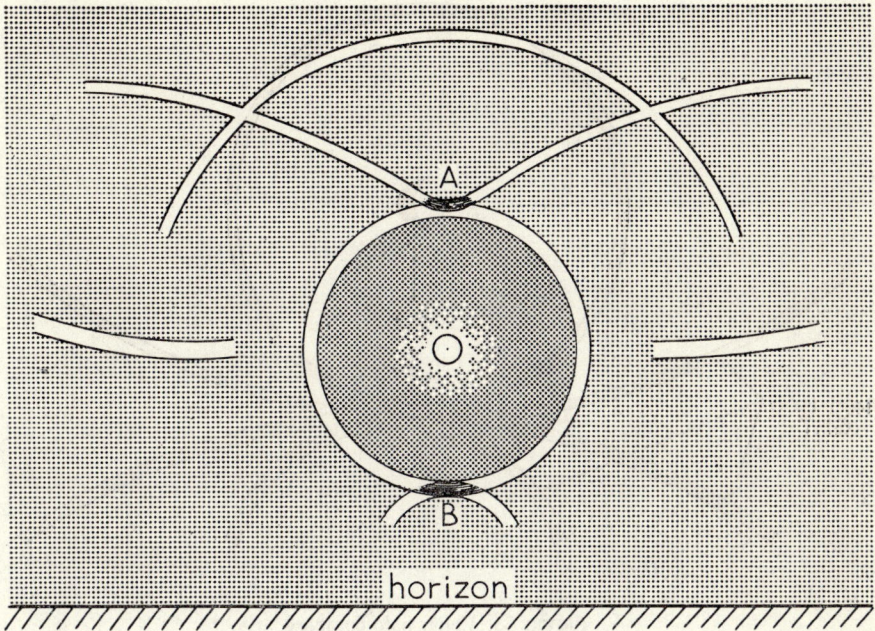

FIGURE 32. Solar halo complex

Witnessed from m.v. *Nova Scotia*, Captain N. R. Land, St John (N.B.) to Liverpool. Observer, Mr A. C. Herdan, 3rd Officer.

'1 October 1965. Position 43°16′N, 66°15′W. The halo complex was clearly seen from 1200–1600 GMT. The radius of the inner halo was 21°26′ and that of the partial outer concentric halo was 46°30′. Two wing-shaped arcs, each subtending an angle of about 54′, crossed the outer concentric halo, meeting at the centre of their span directly above the sun and in contact with the upper edge of the inner concentric halo. At the point of contact (A) a brilliant spectrum could be seen subtending an angle of at least 1¼°. About 1530, the most vivid period, two more haloes were seen. One was a white arc which would have stretched right across the halo complex, passing through the position of the sun, if it hadn't been rendered invisible by the glare. The other was a small, but vivid, inverted half halo of about 6° radius; it was in contact with the inner halo, its upper edge crossing the latter's lower limb (B). At this point also vivid coloration was seen. Altitude of sun: 21°50′; bearing 094°. Cloud, small amounts of Cirrus and Altostratus.'

Having established the position, any point of special interest should be noted, such as an exceptional degree of brightness or colour, variations in brightness in different parts of a halo, or a halo appearing elliptical instead of circular, etc. In the case of the rarer phenomena, the fullest possible information should be recorded, preferably accompanied by a sketch, on which all angular measurements are shown. In sketching halo phenomena the size of the sun (or moon) is usually exaggerated, sometimes very greatly. Even in landscape paintings by well-known artists, the same thing usually occurs. The discs of the sun and moon are about half a degree in diameter and therefore only about one-ninetieth of the diameter of the common halo of 22° radius.

There are few other phenomena not included in the previous pages, e.g. various forms of cross, centred on the sun or moon, are occasionally seen

M
105

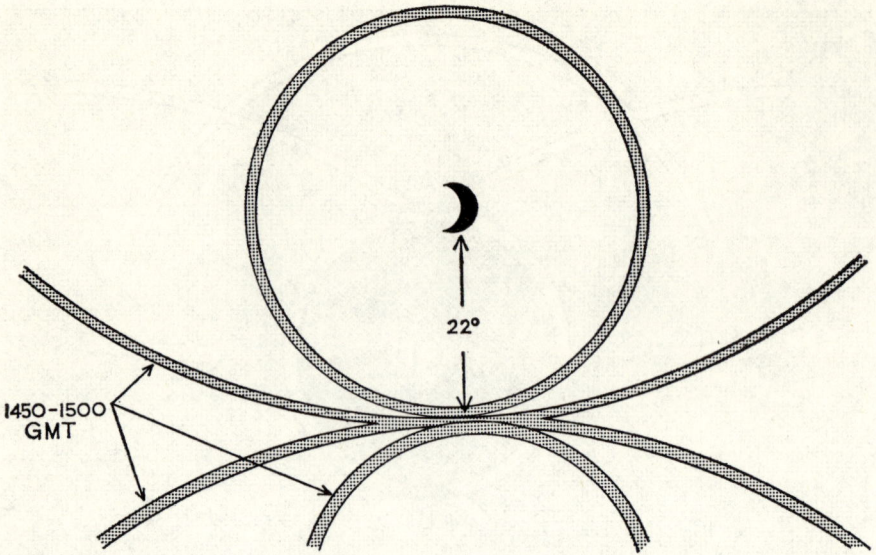

FIGURE 33. Lunar halo complex

Witnessed from m.v. *Koraki*. Captain H. C. Townend. Tauranga to Melbourne. Observer, Mr M. J. C. Orr, 2nd Officer.

'28 May 1967. Position 34°34'S: 171°42'E. The 22° halo was first seen at 1415 GMT when the moon was at an approximate altitude of 60° and bore 090°. At 1450 the tangential arcs appeared, weak at first but gradually gaining in intensity in the next 10 minutes. The complex remained visible for about 20 minutes, after which time only the 22° halo was seen. A thin layer of Cirrostratus to Altostratus was present and also some Cumulus'.

(Figure 34). The vertical arm is usually formed by part of a sun pillar and the horizontal arm by a short portion of the mock-sun circle. Abnormal phenomena of unknown origin are also sometimes reported. One such observation is shown in Figure 32 in which is seen the ordinary 22° halo, with an arc above, probably part of the 46° halo.

IRIDESCENT CLOUD

Patches of delicate, but often vivid, colouring are occasionally seen at any time of the day on altocumulus and other middle and high clouds, often covering quite a large extent of cloud. It may form a very beautiful spectacle, especially if the sun is hidden from the observer's view by lower cloud. Red and green are the most common colours, but others, such as lilac, may be seen. Sometimes the colours lie in bands parallel to the edge of the cloud, but often they form an irregular mosaic, delicately shading into one another. The colouring resembles that of coronae, but the bands of colour do not form concentric circles with the sun as centre. Sometimes a number of coloured patches may be seen along a straight line passing through the sun.

Iridescence is usually seen on cloud near the sun or within about 30° of the sun, but may occur at greater distances. It seems to be most frequently

106

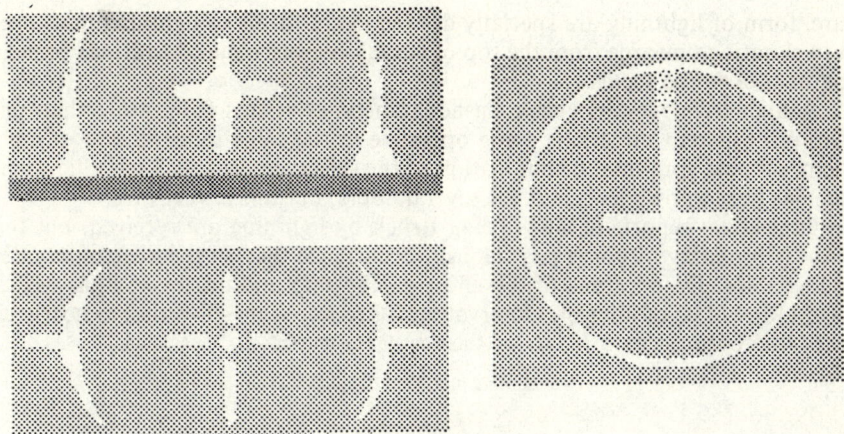

FIGURE 34. Forms of halo cross

observed on cloud that is in process of either formation or evaporation. The colouring is not normally seen after sunset or before sunrise, but brilliant iridescence, continuing after sunset, or appearing before sunrise, may be seen on a very rare high form of cloud, see Mother-of-pearl cloud (page 86).

If the observer is in doubt as to whether he is seeing ordinary sunset cloud colouring, or iridescent colouring towards the time of sunset, it should be remembered that the former may cover large areas of cloud, or many isolated clouds, with one colour. Iridescence, on the other hand, usually shows much smaller areas of different colour on one cloud and the coloration is purer and more prismatic, in this respect resembling the colours of the rainbow.

When seen, remarks on the nature and extent of the colouring, the type of cloud and the approximate angular distance from the sun will be useful.

LIGHTNING

Anything unusual observed during a thunderstorm is worth recording. Some points in connection with lightning are given below.

Lightning varies in colour on different occasions; it is normally white, with perhaps a bluish tinge. Sometimes it is quite a bright violet. Other colours seen are reddish-white, yellowish-white, mauve and blue.

Variations of the ordinary appearance of forked lightning have been seen:

(a) Inequalities of brightness in different parts of the path, known as chain, or beaded lightning, from the impression left on the eye.
(b) Rocket lightning, so called from the relative slowness of the flash, so that the progressive lengthening of the streak can be seen.

The special form known as ball lightning resembles a ball of fire, either falling from a cloud or moving more or less horizontally. It usually lasts only a few seconds and may disappear noiselessly or with an abrupt clap of thunder. Ball lightning has been seen at close range and it has sometimes passed into or through a building. Careful observations of this uncommon, but not extremely

107

rare, form of lightning are specially desired. Rare forms of lightning have been seen shooting upwards from the top of cumulonimbus cloud, in various branching or rocket-like forms.

A high frequency of visible flashes sometimes results from more than one storm in different directions being operative at the same time, so that at night there is amost continuous illumination of the sea or landscape. Such a lightning rate has been known to persist for several hours, but this is very rare.

Occasional reports of ships being struck by lightning are received, but this event is probably of much less frequent occurrence than in the days of wooden sailing ships. Descriptions of the effect on the ship and on the compasses (see page 118) will be of interest. Observations of recent years show that in nearly all cases the foremast or forepart of the vessel is struck.

RAINBOWS

Solar rainbows. The normal appearance of a bright rainbow is as follows. The chief or primary bow shows the sequence of colours, violet, indigo, blue, green, yellow, orange and red, the red being on the outside or top of the bow. In contact with the inside of this bow, one or two fainter 'supernumerary bows' can frequently be seen with the colours in the same order, the first inner bow being much fainter than the primary bow and the second fainter still. Supernumerary bows do not, however, show the full range of spectrum colours; they are essentially red, or red and green, though other colours may be seen. In cases of exceptionally brilliant rainbows up to five supernumerary bows may be seen.

Concentric with the primary bow, but 9° outside it, is the secondary rainbow, in which the full range of colours appear in the reverse order, red inside and violet at the top or outside. The primary bow is formed by means of one internal reflection in each raindrop; the secondary bow is fainter, being produced by two such reflections. The sky between the primary and secondary bow is rather darker than that inside the primary bow, or the general sky in the neighbourhood. The secondary bow is commonly seen, but if the primary bow is faint the secondary one may not be visible.

Both the primary and secondary bows are seen when the observer has his back towards the sun. The sun, the observer's eye and the centres of the circles of which the primary and secondary rainbows form arcs, are always in a straight line, so that the azimuth of the highest part of the bow is 180° from the sun's azimuth. The normal radius of the arc of red light of the primary rainbow is 42°, of the violet arc $40\frac{1}{4}$°; in the secondary bow the radii are 51° for red light and 54° for violet light, all the values given being approximate. Hence the normal breadth of the primary bow is about $1\frac{3}{4}$° and that of the secondary bow about 3°. It also follows that with the sun at an altitude of 42° the uppermost point of the primary bow is on the horizon, its centre being 42° below the horizon, and hence no primary bow can be formed if the sun's altitude exceeds 42°. Similarly no secondary rainbow can be formed if the sun's altitude exceeds 54°. Consequently rainbows are mainly morning and evening phenomena; nearer midday, if seen at all, the arc of the bow is shorter and the altitude small. Thunderstorm rain passing away from the observer gives the most favourable circumstances for the production of bright rainbows.

When the observer is at ground level and the rain cloud is distant, the rainbow arcs are always less than semicircles, unless the sun is on the horizon, when

108

they form semicircles. When, however, the rain is near, and especially if the observer is in an elevated position, such as on the bridge of a ship, the bows will be greater than a semicircle and may even form complete circles. Several accounts have been received of bows forming complete circles as far as the water-line on each side of the ship.

One of the halo phenomena, the circumzenithal arc, may show bright rainbow colouring, but is always in such a position that the observer must face the sun to see it.

Rainbows do not always show the same colouring. The colours seen, and their relative width and intensity, vary according to the size of the raindrops producing the bow. The colours are most brilliant and best defined with very large raindrops such as occur in thunderstorm rain. With fairly large drops, vivid violet and green may be seen, and also pure red, but little or no blue. With smaller drops the red weakens and with still smaller ones the green goes, leaving only the violet. Just before sunset, when the sun is red in colour, especially in autumn and winter, an all-red rainbow may be produced.

If the raindrops are extremely small, as in the case in some cloud and in fog, a white rainbow may be formed. Such a bow is called a 'fog-bow' or 'Ulloa's Ring'. In all rainbows there is some overlapping of the colours; in a white rainbow the overlapping is so complete that white light is reconstituted. For a white rainbow to be seen, the observer must be near the cloud or near or in the fog.

Lunar rainbows. Lunar rainbows are formed in the same way as solar ones, but are considerably rarer, having regard to the comparatively short periods that a bright moon is above the horizon. A lunar rainbow is usually fainter than a solar one and it is not always possible to distinguish colour; the appearance is then whitish. Quite frequently, however, colour may be observed; more rarely the whole sequence of colour can be seen. Secondary and supernumerary lunar rainbows are very rarely seen, on account of their faintness.

Reflection rainbows. These are seen occasionally on calm days when a sheet of water lies in front of or behind the observer standing with his back to the sun. Such bows are formed by rays of light illuminating the falling raindrops after reflection at the surface of the sheet of water. The centre of a reflection rainbow is thus as high above the horizon as the sun, or the same angular distance above, as the centre of the direct bow is below the horizon; consequently the arc, when complete, exceeds a semicircle. The direct and reflection rainbows intersect on the horizon, and the colours have the same sequence.

Observation of rainbows. The observer who wishes to make useful observations of normal rainbows should record the colours seen, in sequence, with an indication of their relative widths and intensities. If supernumerary bows are seen below the primary bow, the number of these and their colouring should be noted. If the secondary bow is unusually bright it is worth while looking for supernumerary bows just above it; these have rarely been seen on account of their faintness. An additional primary bow may be seen when the sea is sufficiently calm to give a reflected image of the sun in the sea, which acts as the light source for the bow. The position of this bow with regard to that formed by the sun itself varies with the sun's altitude. The secondary bow from the sun's reflected image is almost always too faint to be observable.

Abnormal bows, or arcs of bows, perhaps intersecting the normal bows, and sometimes white in colour, have occasionally been seen and it is of special interest to record these as fully as possible, since no explanation has yet been found for some of them. They sometimes meet the horizon at the same point as one of the normal bows. In such cases the sequence of colour, or the absence of colour should be noted. It is essential to give angular measurements of such bows, in the form of azimuths of the ends of the bow or arc, in which case the sun's azimuth should also be given. If a normal bow is also seen, the difference in azimuth between the points where the normal and abnormal bows meet the horizon will serve to establish the position of the latter. If an abnormal bow is seen concentric with the normal primary or secondary bow the difference of altitude of the bows at their highest point should be given.

SCINTILLATION

Scintillation, or twinkling, is the more or less rapid change of apparent brightness of a star, accompanied also at relatively low altitudes by colour changes. It is due to minor changes in the refractive power of the atmosphere. The amount of twinkling is always greatest towards the horizon and least in the zenith. The general amount varies considerably on different nights, so that at the zenith twinkling may be considerable, slight or entirely absent. Nights without appreciable twinkling towards the horizon are rare. When the changes of brightness are small the fluctuations are slower; in proportion as they are greater they become more rapid.

Colour change is usually shown by stars at altitudes not exceeding 34°; it never occurs at altitudes greater than 51°. The brightest stars, e.g. Sirius, at low altitudes show it most and, in favourable conditions, the changes may be very striking, the star flashing blood-red, emerald-green, bright blue, etc.

Scintillation is also observed in the case of terrestrial lights. The shimmering seen near the ground on a hot day is akin to it.

The bright planets do not usually appear to twinkle, as they have discs of definite size, although these are not visible without optical aid. Each point on the disc twinkles independently of the others, so that on the average the light is steady. The planet Mercury, only seen in twilight and at relatively low altitudes may, however, be seen to twinkle because of the small size of its disc, and, exceptionally, other planets at very low altitudes may exhibit some twinkling.

The relative degree of twinkling in different parts of the world, e.g. in temperate as compared with tropical latitudes, is not very well known and any information bearing on this will be of interest. It is probably greatest in temperate latitudes, which are subject to the passage of depressions.

SKY COLORATION, DAY-TIME

The light of the sky in day-time is due to the illumination of the atmosphere by sunlight. The molecules of air exert a selective action on the colour constituents of sunlight, scattering mainly blue rays towards the observer and letting the others pass away.

Dust is always present in greater or less degree in the atmosphere, and in certain states of weather larger particles are present in the lower part of the atmosphere. The presence of dust tends to weaken the blue of the sky, because each particle reflects the whole of the white light. The greater the number of particles and the larger their size the more the sky becomes whitish-blue. For the same reason the cloudless sky is always whiter near the horizon than at higher altitudes. After heavy rain, such as that due to the passage of a depression, the larger dust particles have been washed out of the air and the sky is often a very deep blue.

The sky is often whitened within the region of smoke pollution from a large town. Natural dust from desert sources, at higher levels in the air, also has the same effect, e.g. the white skies seen in the region of the East Indies in the south-east monsoon, caused by dust from the Australian desert. The dust from great volcanic eruptions may whiten the sky for weeks or months afterwards, over more or less considerable areas of the globe. After the Krakatoa eruption the colour of the sky at various times of the day in equatorial regions was described as white, smoky, yellowish or reddish.

The unclouded sky may also be whitened by what is known as 'optical haze', which also makes distant terrestrial objects indistinct. This occurs on hot days and is the result of innumerable little convective uprisings of air, causing confused and variable refraction of light. The shimmering of terrestrial objects on a hot day also results from the same cause.

The sky may sometimes by covered by a layer of cirrostratus, so thin and uniform as not to be visible as cloud, but sufficient to dim the blueness, giving the sky a milky appearance.

A somewhat dirty green coloration of clear patches in a generally overcast sky is sometimes seen at sea in the day-time, not to be confused with the green coloration of part of the clear twilight sky in the west. This day-time coloration is associated with bad or windy weather, or is considered a prognostic of such weather. Observations of it and of the accompanying or subsequent weather will be welcomed, as it is not yet fully understood. It appears to occur most frequently in the Roaring Forties.

SKY COLORATION AT TWILIGHT

When clouds, particularly middle and upper clouds, occur about the time of sunset or sunrise, or in bright twilight, their coloration is often very beautiful. The cloud colours are mainly shades of orange, rose or red, since the direct sunlight illuminating the cloud has passed through a great length of the lower layers of the atmosphere. Shades of purple are sometimes seen, since a cloud may at the same time be indirectly illuminated by scattered blue light from higher atmospheric levels. On rare occasions colouring of exceptional magnificence occurs.

Colour phenomena also occur in a cloudless sky during the twilight periods. These vary considerably and are best developed in arid or semi-arid land regions. Some of those which occur more commonly everywhere are mentioned here. The Primary Twilight Arch appears after the sun has set, as a bright, but not very sharply defined segment of reddish, or yellowish light, resting on the western horizon. After the sun has set, a pink or purple glow may be seen, covering a considerable part of the western sky, known as the First Purple Light. It reaches its greatest brightness when the sun is about 4° below the horizon, and disappears when it is about 6° below.

At sunset, a steely-blue segment, darker than the rest of the sky, begins to rise from the *eastern* horizon. This is the shadow of the earth thrown by the sun on to the earth's atmosphere. The Earth-shadow is bordered by a narrow band of rose or purple colour, called the Counterglow. The whole rises fairly quickly

111

in altitude, the shadow encroaching on the counterglow and soon obliterating it. With increasing general darkness the edge of the shadow weakens, but may sometimes be traced up to its passage through the zenith. In the later stages of twilight, this shadow edge has come down nearly to the *western* horizon, leaving a slightly more luminous segment between it and the horizon. This is the Secondary Twilight Arch. Just before the ending of astronomical twilight, it is sometimes seen as a fairly well defined whitish arch on the horizon, with an altitude of only a few degrees at its apex. This might be confused with an auroral arc visible at very low altitude.

Analogous phenomena, in the reverse order, occur before sunrise. Other colours are often seen in the cloudless twilight sky, portions of which may be green, yellow, orange or red, according to the amount of dust and water vapour present in the air. Instead of the purple light after sunset, the sky very often shows some shade of clear green, probably when the air is relatively free from dust.

SKY COLORATION AT NIGHT

Between the visible stars, brighter and fainter, the background of the clear night sky is not wholly dark. Part of the general luminosity of the sky is due to the accumulated light of the brighter telescopic stars, which cannot be seen as individual stars with the unaided eye. The remainder is due to the airglow, which may vary in intensity on different nights, see page 91. The airglow is greenish in colour but it is usually too faint for the colour to be seen. In bright moonlight the sky is generally somewhat greenish, but it is probable that when the air is relatively dust-free and the full moon is at high altitude it becomes bluish. Opinion varies on this point, as the colour of faint light is not equally well seen by different persons.

TWILIGHT

Twilight is due to the illumination of the higher levels of the atmosphere by the sun when this is below the observer's horizon. The last stage of twilight is very faint and indefinite and it is not possible to say exactly when it ends. Astronomical twilight is defined as ending in the evening when the sun's centre is 18° below the horizon, since by that time sixth magnitude stars, the faintest that can be seen by the naked eye, have become visible in the region of the zenith.

Another and shorter twilight period, that of civil twilight, is recognized; this ends in the evening when the sun is 6° below the horizon. This is assumed to mark the ending of the time when outdoor labour is possible. The period of civil twilight is important to the seaman because experience has shown that subsequent to it the horizon is not sufficiently clearly visible to obtain good stellar observations. In the later stages of civil twilight such observations can be made, the brighter fixed stars being visible and the horizon still remaining clearly visible. Similar definitions apply to morning twilight.

The duration of twilight varies according to the latitude. It is shortest in the tropics where the apparent track of the sun down to the horizon is steepest. It also varies to some extent at different seasons, being shortest in all latitudes about the time of the equinoxes. The following table shows the extent of these variations between the equator and latitude 60°N or S; AT and CT refer to astronomical and civil twilight respectively.

	Equator		30°		50°		60°	
	AT	CT	AT	CT	AT	CT	AT	CT
	h min	h min	h min	h min	h min	h min	h min	h min
Midwinter ...	1 16	0 26	1 26	0 31	2 1	0 45	2 48	1 9
Equinoxes ...	1 10	0 24	1 20	0 28	1 52	0 37	2 31	0 48
Midsummer ...	1 15	0 26	1 37	0 32	—	0 51	—	1 59

In the belt between latitude $48\frac{1}{2}°$ and the Arctic Circle there is no true night for some weeks of the midsummer period, as the sun does not sink as much as 18° below the horizon. There is a similar belt in the southern hemisphere, six months later, during the southern summer. In polar regions there is a long twilight period of about two months between the long polar periods of summer daylight and winter night.

At rare intervals abnormally long duration of twilight is observed. This is caused by the presence of fine dust suspended in the upper air. The dust may be due to a great volcanic eruption, such as that of Krakatoa in 1883 or to the fall of an exceptionally large meteor, such as that of 30 June 1908 in Siberia. Observations of exceptionally bright and long-continued twilight will be of value.

WATERSPOUTS

A waterspout is a whirlwind over the sea, appearing as a funnel-shaped column usually extending from the lower surface of cumulonimbus cloud to the sea. In travelling over the sea this column often becomes oblique or bent; it may become looped. The spout is in rapid rotation and the wind around it follows a circular path. Although very local, this wind is often violent, causing confused but not high sea. A noise of 'rushing wind' may be heard. A waterspout in most cases forms downwards from the base of the cloud, appearing in its earlier stages as a dark funnel hanging from the cloud. The sea surface below becomes agitated and the funnel finally dips into the centre of the spray. The waterspout may last from a few minutes up to half an hour or more. Sometimes the spout, formed of condensed water vapour, does not reach the sea, and retreats up into the cloud. Several may be seen at the same time.

There are a number of theories which attempt to explain the formation of a waterspout. These theories may be classified into those which relate to the origin of the more severe tornado storm spout of the tropics and subtropics and those which relate to the milder 'fair weather' spout of the tropical and temperate latitudes.

The tornado spout may form over the sea but is more likely to have formed over land and subsequently to have passed out to sea. Its formation may result from the horizontal shear between warm and cold air currents existing up to considerable heights in the atmosphere. Such conditions normally occur along a cold front or cold occlusion surface. The tornado storm waterspout may damage even a large vessel if it passes directly over it, the damage being caused

113

partly by tornadic winds, partly by the suddenly reduced pressure and partly by the deluge of water sometimes released.

The 'fair weather' waterspout is believed to be formed mainly by convection processes. Under conditions of a high temperature lapse rate near the sea surface, a small parcel of moist air becomes a little warmer than its environment and begins to rise. Rotation is caused by the converging surface winds sucked in under the rising air parcel and energy gained by the atmospheric instability is augmented by the latent heat of condensation of the water vapour present. The initial convectional ascending air current may occur directly below cumulonimbus cloud in which case it may penetrate the cloud, and the rotation increases until a complete waterspout is formed. Although the 'fair weather' waterspout should cause no real damage to a large vessel, it should be avoided by the small-boat mariner.

Observations of waterspouts, with sketches or photographs, and details of their mode of formation and dissipation are of value. The diameter of the spout and the direction of rotation should be noted. If it is possible to determine the rate of rotation, this information is very valuable. Sometimes a streak or mark on the spout enables this to be done. The spout is a hollow tube; double-walled spouts have occasionally been recorded. The approximate vertical height of a spout may be found by sextant measurement of the angle subtended, together with the known or estimated distance from the ship. The height of a waterspout from sea surface to cloud base is usually from 1000 feet (300 m) to 2000 feet (600 m). It may, however, be as little as 100 feet (30 m) or as much as 5000 feet (1500 m). There is a very great variation in the observed diameters of waterspouts, from 1 to about 200 metres.

Though waterspouts are infrequent in high latitudes their frequency does not depend wholly on latitude. In general, more are observed in lower latitudes but their frequency in tropical and equatorial regions varies considerably in different oceans. Waterspouts are most common in the following regions: the Equatorial Atlantic; the South Atlantic; the eastern coast of the United States, south of lat. 35°N; the Gulf of Mexico; part of the eastern Mediterranean; the Bay of Bengal; the Gulf of Siam.

CHAPTER 12

Marine Phenomena

SEA COLORATION

The normal colour of the sea in the open ocean in middle and low latitudes is an intense blue or ultramarine. The following modifications occur elsewhere:

(a) In all coastal regions and in the open sea in higher latitudes, where the minute floating animal and vegetable life of the sea, called plankton, is in greater abundance, the blue of the sea is modified to shades of bluish-green and green. This results from a soluble yellow pigment, given off by the plant constituents of the plankton.

(b) When the plankton is very dense, the colour of the organisms themselves may discolour the sea, giving it a more or less intense brown or red colour. The Red Sea, Gulf of California, the region of the Peru Current, South African waters and the Malabar Coast of India are particularly liable to this, seasonally.

(c) The plankton is sometimes killed more or less suddenly, by changes of sea temperature etc., producing dirty-brown or grey-brown discoloration and 'stinking water'. This occurs on a unusually extensive scale at times off the Peruvian coast, where the phenomenon is called 'Aguaje'.

(d) Larger masses of animate matter, such as fish spawn or floating kelp, may produce other kinds of temporary discoloration.

(e) Mud brought down by rivers produces discoloration, which in the case of the great rivers may affect a large sea area. Soil or sand particles may be carried out to sea by wind or duststorms, and volcanic dust may fall over a sea area. In all such cases the water is more or less muddy in appearance. Submarine earthquakes may also produce mud or sand discoloration in relatively shallow water, and oil has sometimes been seen to gush up. The sea may be extensively covered with floating pumice stone after a volcanic eruption.

It is desirable to record all cases of unusual sea coloration. To determine the cause, microscopic examination of a sample may be necessary, and whenever possible a sample should be taken for subsequent examination at the Institute of Oceanographic Sciences. The sample can be preserved for a considerable time if a few drops of 40 per cent formalin or of a strong solution of mercuric chloride are added. Port Meteorological Officers in UK ports carry sets of bottles and preservative for this purpose and will supply any shipmaster on request.

ABNORMAL RISES OF SEA LEVEL AND ABNORMAL WAVES

Both these phenomena are popularly included in the term 'tidal waves', but neither has any connection with the tides. If either occurs, however, at a coast in conjunction with a high tide, its effect will obviously be greater and more destructive.

115

Abnormal rises of sea level, on which ordinary sea and swell waves are superimposed, are produced by severe storms. High water levels are thus caused on many coasts, but fortunately the rise is rarely large enough to cause great damage. With strong westerly winds the water level at Cuxhaven, at the mouth of the Elbe, may rise $2\frac{1}{2}$ metres above the normal. On exceptional occasions the rise has reached $3\frac{1}{2}$ metres above the normal. Destructive rises mainly occur in connection with tropical revolving storms; rises of as much as 6 and $4\frac{1}{2}$ metres have been experienced on parts of the east coast of the United States.

Submarine earthquakes and landslides, and violent volcanic eruptions near a coast or on an island, produce abnormal waves. Sometimes these are visible waves, at other times shock waves, the latter giving the sensation in severe cases of the ship having struck a rock. The visible waves may travel many hundreds of miles, or in very severe disturbances many thousands of miles.

Single high waves in fair weather, with smooth or moderate sea, are almost certainly of seismic origin. Sometimes there may be two or more such waves at intervals. On the other hand, isolated giant waves which have been reported in gale conditions, are probably caused by a synchronism of the larger waves in a sea or swell cycle. Some of these have been estimated to reach or exceed a height of 18 metres.

There have been, in recent years, a number of reports of abnormal waves causing considerable damage to quite large ships. There have also been reports of smaller vessels lost without trace possibly as a result of the action of such waves. It has thus become recognized that from time to time, particularly in certain sea areas, there can occur very unusual or 'freak' waves, the causes of which are not yet fully understood and concerning whose frequency we do not know nearly enough.

A 'freak' wave has recently been defined as a wave of very considerable height ahead of which there is a deep trough, so that it is the steepness of the wave which is its outstanding feature and which makes it dangerous to shipping. Many of the reports of 'freak' waves have come from an area off the coast of south-east Africa during the period May to October. It is thought to be very significant that this is an area where a strong current (the Aghulas Current) runs counter to the high seas generated by the rather frequent south-westerly gales of the winter months and also to the unusually heavy swells which spread north-eastwards from the Southern Ocean at that time of the year. Theory indicates that a counter current opposes the advance of the wave energy through a sea and that when the current speed reaches one-quarter of the speed of the waves the wave energy will be trapped, leading to an area of steep and confused waves beyond which there is a patch of relatively calm water. In practice this is not inconsistent with the very occasional occurrence of a very high wave, produced by sea and swell waves getting into phase, whose front is much steepened by an opposing current.

There are probably other ocean areas where conditions favourable for these 'freak' waves occur from time to time, e.g. in the vicinity of the Gulf Stream in a period of north-easterly gale. More information is greatly needed. Whenever these abnormal wave conditions are met with they should be reported in as much detail as possible. Besides the exact time and position, weather conditions and the course and speed of the ship, information is needed about wind and wave conditions, both before and after the encounter, about any other factors which may influence the state of the sea, and of course a full description of the

116

'freak' wave itself together with a brief note about any damage sustained. This information should be entered in the 'Additional remarks' pages of the meteorological logbook or on a Freak Wave Report form available from Port Meteorological Offices.

MARINE BIOLUMINESCENCE

This phenomenon exhibits many different forms. The more remarkable of these include:

(a) the diffused white light, which may give enough light to read by, or to illuminate clouds: it is called 'white water' or 'milky sea'. The even glow is believed to be due to light from marine organisms of microscopic size. It is especially prevalent in the Arabian Sea;

(b) systems of moving parallel bands;

(c) rapid light flashes on the sea surface;

(d) upwelling of subsurface water, breaking into vivid luminosity at the surface;

(e) the great systems of bands rotating round a central luminous 'hub', known as 'phosphorescent wheels'.

The last-named is a well-authenticated but, in general, rather infrequent phenomenon, apparently confined to the Indian Ocean north of the equator and the China Sea region. One or more wheels may occur simultaneously, rotating in the same or opposite directions.

While progress has been made in classifying the varied forms of luminescence, it is not yet possible to explain how the majority of them are caused, either as regards the animal organisms involved or the nature of the stimulus producing light-emission. Many of the apparent movements of luminous areas, e.g. those of the 'spokes' of the wheel, are much too rapid to be caused by the actual movements of organisms through the water. The most inexplicable phenomenon is that of luminescence in the air a few feet above the sea surface when there is no light in the water; this has been reported on several occasions.

Most of our knowledge of the varied forms of marine bioluminescence has been derived from the observations recorded in ships' logbooks and continued observations will be of the greatest value. When sufficient information is accumulated marine biologists will be able to work on the problems connected with the causation of the various phenomena. Observations should be as precise and detailed as possible and should include estimations of the direction, length and width of moving bands and of the size of what appear to be individual luminous organisms. A water sample would be useful if it could be treated with a preservative. In any case a record of whether a water sample is luminous when stirred or shaken would be very useful, together with the appearance of any such luminescence. In the case of moving or rotating bands a careful estimate of the time interval between the passage of successive bands should be made. If a wheel is seen it is important to note if it has a visible centre and to estimate the distance of the centre from the ship, also to record the direction of rotation of the wheel. The wheel sometimes forms from parallel moving bands, or changes into these, and all changes of form of the luminescence should be recorded.

117

Interesting experiments on the possible nature of the stimulus may be made:

(a) By trying the effect of flashing light on the sea, e.g. from an Aldis light; sometimes this initiates or increases luminescence, but not always.
(b) The effect of the switching on and off of radar.
(c) No observation of a wheel recorded by a sailing vessel has been found and it is therefore possible that it is produced by sound or vibration waves through the water from a modern vessel. It would be of great value to note the effect on the wheel, if any, produced by stopping the engines for a few minutes.

The luminescence usually appears white, in the case of 'white water', and various shades of blue and green, in other forms. Other colours have, however, been recorded. In all observations it is important to give the colour; in the case of the more striking moving forms, observers apparently find them so impressive in other ways that this important factor is almost invariably overlooked. There is thus very little information, for example, about the colour of the light of the wheel.

ABNORMAL COMPASS DEVIATIONS

A ship's magnetic compass may show appreciable deviation during the progress of a considerable or severe magnetic storm (see Magnetic Disturbances, page 84).

When an aurora of an active type is seen, especially in latitudes lower than those in which aurora is normally seen, the possibility of deflections of the magnetic compass should always be borne in mind. Mere brightness of aurora in a region where aurora frequently occurs is no criterion of the occurrence of a magnetic storm, e.g. a bright, colourless and relatively quiescent aurora seen in August or September in the western Atlantic on the Belle Isle route.

If a ship be struck by lightning, a sudden abnormal deviation of the compass may result. This error may be of a temporary or a permanent nature. Chronometers may also be affected.

Abnormal magnetic variation occurs locally in various regions. These variations, if experienced, should always be recorded, particularly if no mention of abnormal variation is made in the appropriate Admiralty *Pilot* or on Admiralty charts of the region.

Part IV   Summary of Meteorological Work at Sea

CHAPTER 13

Organization of Voluntary Meteorological Work at Sea

Historical. M. F. Maury, an officer of the US Navy, was the first man to realize the commercial and scientific value of weather information collected from ships. Owing to his initiative, the first International Meteorological Conference was held at Brussels in 1853 to consider international co-operation and a uniform system of observation. Following this conference, the British Meteorological Office was established in 1854, under Admiral FitzRoy, as a Department of the Board of Trade.

On assuming office, FitzRoy issued a circular letter to the masters of merchant ships, inviting their co-operation in observing the weather at sea, and by 1855, 105 ships of the Mercantile Marine and 32 ships of the Royal Navy were equipped with instruments for this purpose.

Observations were originally recorded in a 'Weather Register' whose general form was agreed upon at the Brussels Conference. In 1874, Captain Henry Toynbee, who had then been Marine Superintendent of the Meteorological Office for seven years, drew up a 'Meteorological Log' based on the original Weather Register, but incorporating improvements. This was approved internationally and brought into use by the Meteorological Office for British ships. This Log has been the means of providing climatological atlases for all oceans, and has provided a basis for scientific investigation. It underwent very little change up to the end of World War I, when the use of climatological logbooks was gradually discontinued in favour of observations made at synoptic hours and transmitted by radio. In 1953 the method of setting out the observations was entirely rearranged to produce the present-day meteorological logbook which is a combined record of observations made and radio weather messages sent.

In 1861, FitzRoy instituted the system whereby certain ports were informed by telegraph of impending gales and were asked to hoist visual gale-warning signals for the benefit of shipping. Except for a short break in 1867–68, this system has been maintained up to the present day.

The invention of wireless telegraphy opened up a new era in marine meteorology. As early as 1906, HM ships sent observations to the Meteorological Office by radio, while in 1909 a number of transatlantic liners commenced a similar service of reports by radio. Owing to the disruption caused by World War I, it was not until 1921, as a result of arrangements made by the International Meteorological Organization, that radio weather messages from merchant ships were organized on a satisfactory scale, and an international code was introduced for the purpose. In this year a number of Selected Ships commenced not only recording their observations, but transmitting them by radio in a special code at the internationally agreed hours of 0000, 0600, 1200 and 1800 GMT. These messages were sent from all oceans through designated shore radio stations to various meteorological centres in accordance with an international scheme. The number of ships which continued merely to record their observations six times daily, at the end of each watch, was gradually reduced as

119

the number recording and reporting at the synoptic hour was increased. Today all observing ships report at the main synoptic hours and the data recorded in this manner can still be used for climatological purposes.

At a meeting of the International Convention for Safety of Life at Sea, held in 1929, provision was made (in Article 35) for the international encouragement of meteorological work at sea. This Convention was revised in 1948 and again in 1960 (see Chapter 14).

During World War II observations from merchant ships again ceased. In 1946, as a result of a conference held in London, it was agreed that all meteorological services of the British Commonwealth would co-operate in organizing meteorological work at sea. In 1947 the International Meteorological Organization introduced a new universal code for the sending of radio weather messages by voluntary observing ships of all nations.

Throughout the history of the Marine Division of the Meteorological Office, observations at sea have been made on a voluntary basis. The number of ships making observations at any time depends upon requirements but is limited by practical considerations. The masters and officers of ships undertaking this work are referred to as the 'Corps of Voluntary Marine Observers', their ships comprising the 'Voluntary Observing Fleet'.

Marine meteorology and the British Commonwealth. The arrangements for this generally follow those organized and kept up to date by the World Meteorological Organization (see page 121).

Voluntary Observing Ships of all nations are divided into three main classes: 'Selected Ships', 'Supplementary Ships' and 'Auxiliary Ships'. The first-named observe wind, weather, pressure and barometric tendency, temperatures, clouds and waves. They are equipped with a marine mercurial barometer fitted with a correction slide or precision aneroid barometer Mk II, a barograph, wet- and dry-bulb thermometers in a modified marine screen, and a sea thermometer and bucket.

Supplementary Ships make the same observations with the exception of barometric tendency, sea temperature and waves, and are not therefore equipped with barograph, bucket or sea thermometer.

Auxiliary Ships make similar observations to those made by Supplementary Ships, except that they do not report cloud. They use their own instruments, which have been previously checked by a Port Meteorological Officer to observe pressure and temperature. They only record and report when in areas where shipping is normally sparse.

In addition to the above, many ships engaged in the coastwise and short-sea trades around the British Isles are supplied with a sea-temperature bucket and thermometers. They radio sea-water temperatures to the British Meteorological Office. Several ships trading across the North Sea, and trawlers operating in distant waters, are asked to radio reports of wind and weather, not involving the use of instruments, at synoptic hours whenever possible. A number of distant-water trawlers, however, with a special aptitude for the work are recruited as Supplementary Ships.

Each Meteorological Service is responsible for recruiting its own ships. In addition, each Service may recruit ships of other registries which sail regularly from its ports and do not return to home ports for long periods.

Representatives of Commonwealth and foreign Meteorological Services may, if they wish, visit any British Selected or Supplementary Ship to discuss local problems, supply forms, maps or local information, attend instruments, take extracts from logbooks or express appreciation of services rendered. They may also, if the situation warrants (i.e. paucity of observations in their area), visit other British ships and request their co-operation as Auxiliary Ships, when in the area of the Service concerned. Commonwealth Services inform the British Meteorological Office of the names of Selected and Supplementary Ships recruited by them. The British Meteorological Office promulgates this information, together with the names of all ships recruited in Britain, in each July number of *The Marine Observer*.

The World Meteorological Organization. While the Commonwealth Conference provides for uniformity of practice in marine meteorology among the various Services of the British Commonwealth, the World Meteorological Organization performs a similar function internationally.

Meteorology is so international in character that co-operation is necessary between all countries of the world. This was recognized as long ago as 1872 when the International Meteorological Organization (IMO) was formed, which has ever since acted as an advisory body to National Meteorological Services, its primary functions being the standardization of codes and procedure, the improvement of meteorological practice, and the promotion of research. The Selected Ship Scheme and the issue of weather bulletins for shipping on a world-wide basis, are co-ordinated in this way.

The IMO was a semi-official body, and in 1947 it was decided that, in view of the growing world importance of meteorology for commercial, economic and scientific purposes, it was necessary to change the status of this organization. As a result an intergovernmental body, the World Meteorological Organization (WMO), held its first congress in Paris during 1951, and took over the duties and responsibilities of the IMO. In this organization technical problems are deliberated by a number of technical commissions, whose members are all experts in their particular sphere. All aspects of maritime meteorology are thus dealt with by the Commission for Marine Meteorology, which advises the WMO as necessary.

The instructions to observers issued by the Marine Division of the Meteorological Office conform to the advice of the WMO. Such changes of codes and procedure as occur from time to time are the result of international agreement. It is inevitable that progress in meteorology should bring changes of procedure. Such changes are kept to a minimum, and the basic aim is that every change should achieve greater world-wide application and uniformity, and hence simplicity.

All meteorological work done by ships' officers is entirely voluntary. Only by a voluntary scheme can the requisite high standard of observations be maintained. The benefit of this work to mariners lies in the fact that it forms the basis of the mateorological services for shipping outlined below.

Voluntary observing ships are requested to report their observations at the standard synoptic hours, viz. 0000, 0600, 1200, 1800 GMT, using the standard International Weather Code, either in full or abbreviated form. Information regarding this code and full instructions for coding are to be found in the *Ships' Code and Decode Book* (Met. O. 509) or in the *Admiralty List of Radio Signals*, Vol. 3.

Meteorological services for shipping. The first meteorological service for shipping was the issue of visual Gale Warnings, started in 1861. In 1924, a Radio Weather Shipping Bulletin was instituted; this contained weather reports from certain coastal stations and forecasts for areas around the British Isles. Since

121

then, meteorological warnings of all kinds have been broadcast direct to shipping by radio on an international basis, under arrangements made by the World Meteorological Organization.

Present-day weather messages to shipping aim at providing not only forecasts but such basic information as will enable simple synoptic charts to be drawn on board ship. Such messages are generally known as 'bulletins'. They usually contain:

A brief statement of the meteorological situation.
Area forecasts.
Land station reports.
Ships' reports.
Analyses in the International Analysis Code (I.A.C. (Fleet)).

An example of such a bulletin is the Atlantic Weather Bulletin, full particulars of which are given in Met. O. 509, *Ships' Code and Decode Book*, and in the *Admiralty List of Radio Signals*, Vol. 3. Briefer bulletins for the benefit of coastal shipping are issued by Post Office coastal stations on W/T and R/T and by the BBC (see *Met. O. Leaflet No. 3.* obtainable free from the Meteorological Office).

A valuable facility now provided by many meteorological services consists of the broadcasting of weather maps, both actual and forecast, by means of radio facsimile. By this means complete weather maps drawn by meteorologists ashore can be reproduced on board ship with minimum delay. Not only weather maps but also wave height and sea-ice distribution maps can be received in this way. Installation of the appropriate receiving equipment is, of course, necessary but up-to-date information then becomes readily available with the minimum of trouble. (See also *Meteorology for Mariners*.)

The experience of generations of observers is available in the vast number of observations from the sea that have been collected since 1854. The task of the Marine Division of the Meteorological Office has been not only to collect these observations but to classify and analyse them scientifically and to prepare climatological and other material based upon them, for the information of mariners, and of the world in general. The observations, being (except for those from weather ships) the only ones available from the oceans, are put to many other useful purposes, and are of great value for research into meteorological problems. Most of the analysis is carried out with the aid of computers, and the final results, after careful scrutiny by climatological experts, are issued in the form of atlases for the different oceans. The atlases contain means values for each month of the various meteorological elements observed at sea, and enable the user to assess average conditions at any time in almost any part of the world.

Guidance for conduct of the work at sea. Direct contact between the Meteorological Office and ships' masters and observers is maintained through Port Meteorological Officers at Cardiff, Glasgow, Liverpool, Hull, London, Newcastle and Southampton.

Indirect contact with the Observing Fleet is maintained through the medium of *The Marine Observer*, a quarterly publication which contains articles on meteorology, oceanography, ice, etc., of interest to seamen. A large section in each number is devoted to observations of phenomena of a meteorological or general scientific nature, mostly extracted from the meteorological logbooks of ships of the British Commonwealth.

Instruments are supplied to ships by Port Meteorological Officers and are delivered by hand. When it is desired to return instruments lent by the Meteorological Office, the appropriate Port Meteorological Officer should be advised. When this is not possible, as for example at certain small ports, application should be made to the Marine Superintendent of the Meteorological Office for instructions. Similar remarks apply to the return of damaged instruments for repair or replacement.

Any accident to an instrument, even though no apparent damage is done, should be reported to the Port Meteorological Officer. This is necessary because the constants of the instrument may have been altered without any apparent difference in its working. On no account should a barometer or any other instrument belonging to the Meteorological Office be sent to an instrument maker for repair, or any attempt be made to repair the instrument on board the ship.

CHAPTER 14

The International Conference on Safety of Life at Sea, 1960*

Upon the invitation of the Inter-Governmental Maritime Consultative Organization, a Conference was held in London from 17 May to 17 June 1960 for the purpose of drawing up a Convention to replace the International Convention for the Safety of Life at Sea signed in London on 10 June 1948 as well as for the purpose of revising the International Regulations for Preventing Collisions at Sea, 1948.

This Conference took cognizance of the mariners' requirements for meteorological information and their ability to detect and warn others of hazardous conditions. The following regulations were therefore included in the SOLAS Convention, Chapter V—Safety of Navigation.

Regulation 2: *Danger Messages*

(a) The master of every ship which meets with dangerous ice, a dangerous derelict, or any other direct danger to navigation, or a tropical storm, or encounters sub-freezing air temperatures associated with gale force winds causing severe ice accretion on superstructures, or winds of force 10 or above on the Beaufort scale for which no storm warning has been received is bound to communicate the information by all the means at his disposal to ships in the vicinity, and also to the competent authorities at the first point on the coast with which he can communicate. The form in which the information is sent is not obligatory. It may be transmitted either in plain language (preferably English) or by means of the International Code of Signals. It should be broadcast to all ships in the vicinity and sent to the first point on the coast to which communication can be made, with a request that it be transmitted to the appropriate authorities.

(b) Each Contracting Government will take all steps necessary to ensure that when intelligence of any of the dangers specified in paragraph (a) is received, it will be promptly brought to the knowledge of those concerned and communicated to other interested Governments.

(c) The transmission of messages respecting the dangers specified is free of cost to the ships concerned.

(d) All radio messages issued under paragraph (a) of this Regulation shall be preceded by the Safety Signal, using the procedure as prescribed by the Radio Regulations as defined in Regulation 2 of Chapter IV.

Regulation 3: *Information required in Danger Messages*

The following information is required in danger messages:

---

*See *International Conference on Safety of Life at Sea*, 1960, H.M.S.O., London.

(a) Ice, Derelicts and other Direct Dangers to Navigation.
  (i) the kind of ice, derelict or danger observed;
  (ii) the position of the ice, derelict or danger when last observed;
  (iii) the time and date (Greenwich Mean Time) when danger last observed.

(b) Tropical Storms—(Hurricanes in the West Indies, Typhoons in the China Sea, Cyclones in Indian waters, and storms of a similar nature in other regions).
  (i) A statement that a tropical storm has been encountered. This obligation should be interpreted in a broad spirit, and information transmitted whenever the master has a good reason to believe that a tropical storm is developing or exists in his neighbourhood.
  (ii) Time, date (Greenwich Mean Time) and position of ship when the observation was taken.
  (iii) As much of the following information as is practicable should be included in the message:
    —barometric pressure, preferably corrected (stating, millibars, inches, or millimetres, and whether corrected or uncorrected);
    —barometric tendency (the change in barometric pressure during the past three hours);
    —true wind direction;
    —wind force (Beaufort scale);
    —state of the sea (smooth, moderate, rough, high);
    —swell (slight, moderate, heavy) and the true direction from which it comes. Period or length of swell (short, average, long) would also be of value;
    —true course and speed of ship.

(c) Subsequent Observations. When a master has reported a tropical or other dangerous storm, it is desirable, but not obligatory, that further observations be made and transmitted hourly, if practicable, but in any case at intervals of not more than three hours, so long as the ship remains under the influence of the storm.

(d) Winds of force 10 or above on the Beaufort scale for which no storm warning has been received.

This is intended to deal with storms other than the tropical storms referred to in paragraph (b); when such a storm is encountered, the message should contain similar information to that listed under paragraph (b) but excluding the details concerning sea and swell.

(e) Sub-freezing air temperatures associated with gale force winds causing severe ice accretion on superstructures.
  (i) Time and Date (Greenwich Mean Time).
  (ii) Air temperature.
  (iii) Sea temperature (if practicable).
  (iv) Wind force and direction.

Examples

Ice

TTT Ice. Large berg sighted in 4605 N, 4410 W, at 0800 GMT, May 15.

125

Derelicts

TTT Derelict. Observed derelict almost submerged in 4006 N, 1243 W, at 1630 GMT. April 21.

Danger to Navigation

TTT Navigation. Alpha lightship not on station. 1800 GMT. January 3.

Tropical Storm

TTT Storm. 0030 GMT. August 18. 2204 N, 11354 E. Barometer corrected 994 millibars, tendency down 6 millibars. Wind NW, force 9, heavy squalls. Heavy easterly swell. Course 067, 5 knots.

TTT Storm. Appearances indicate approach of hurricane. 1300 GMT. September 14. 2200 N, 7236 W. Barometer corrected 29·64 inches, tendency down ·015 inches. Wind NE, force 8, frequent rain squalls. Course 035, 9 knots.

TTT Storm. Conditions indicate intense cyclone has formed. 0200 GMT. May 4, 1620 N, 9203 E. Barometer uncorrected 753 millimetres, tendency down 5 millimetres. Wind S by W, force 5. Course 300, 8 knots.

TTT Storm. Typhoon to southeast 0300 GMT. June 12. 1812 N, 12605 E. Barometer falling rapidly. Wind increasing from N.

TTT Storm. Wind force 11, no storm warning received. 0300 GMT. May 4. 4830 N, 30 W. Barometer corrected 983 millibars, tendency down 4 millibars. Wind SW, force 11 veering. Course 260, 6 knots.

Icing

TTT experiencing severe icing. 1400 GMT. March 2. 69 N, 10 W. Air temperature 18 (F). Sea temperature 29 (F). Wind NE, force 8.

Regulation 4: *Meteorological Services*

(a) The Contracting Governments undertake to encourage the collection of meteorological data by ships at sea and to arrange for their examination, dissemination and exchange in the manner most suitable for the purpose of aiding navigation. Administrations shall encourage the use of instruments of a high degree of accuracy, and shall facilitate the checking of such instruments upon request.

(b) In particular, the Contracting Governments undertake to co-operate in carrying out, as far as practicable, the following meteorological arrangements:

  (i) To warn ships of gales, storms and tropical storms, both by the issue of radio messages and by the display of appropriate signals at coastal points.

 (ii) To issue daily, by radio, weather bulletins suitable for shipping, containing data of existing weather, waves and ice, forecasts, and when practicable, sufficient additional information to enable simple weather charts to be prepared at sea and also to encourage the transmission of suitable facsimile weather charts.

(iii) To prepare and issue such publications as may be necessary for the efficient conduct of meteorological work at sea and to arrange, if practicable, for the publication and making available of daily weather charts for the information of departing ships.

 (iv) To arrange for selected ships to be equipped with tested instruments (such as a barometer, a barograph, a psychrometer, and suitable apparatus for measuring sea temperature) for use in this service, and to take meteorological observations at main standard times for surface synoptic observations (at least four times daily, whenever circumstances permit) and to encourage other ships to take observations in a modified form, particularly when in areas where shipping is sparse; these ships to transmit their observations by radio for the benefit of the various official meteorological services, repeating the information for the benefit of ships in the vicinity. When in the vicinity of a tropical storm, or of a suspected tropical storm, ships should be encouraged to take and transmit their observations at more frequent intervals whenever practicable, bearing in mind navigational pre-occupations of ships' officers during storm conditions.

(v) To arrange for the reception and transmission by coast radio stations of weather messages from and to ships. Ships which are unable to communicate direct with shore shall be encouraged to relay their weather messages through ocean weather ships or through other ships which are in contact with shore.

(vi) To encourage all masters to inform ships in the vicinity and also shore stations whenever they experience a wind speed of 50 knots or more (force 10 on the Beaufort scale).

(vii) To endeavour to obtain a uniform procedure in regard to the international meteorological services already specified, and, as far as is practicable, to conform to the Technical Regulations and recommendations made by the World Meteorological Organization, to which the Contracting Governments may refer for study and advice any meteorological question which may arise in carrying out the present Convention.

(c) The information provided for in this Regulation shall be furnished in form for transmission and transmitted in the order of priority prescribed by the Radio Regulations, and during transmission 'to all stations' of meteorological information, forecasts and warnings, all ship stations must conform to the provisions of the Radio Regulations.

(d) Forecasts, warnings, synoptic and other meteorological reports intended for ships shall be issued and disseminated by the national service in the best position to serve various zones and areas, in accordance with mutual arrangements made by the Contracting Governments concerned.

TABLE 1
TEMPERATURE CORRECTION OF THE MET. O. KEW-PATTERN BAROMETER MK 1
(Inch Scale)

To be used with barometers having National Physical Laboratory certificate dated ON OR BEFORE 31 DECEMBER 1954.

These corrections are to be **subtracted** *from the barometer readings to reduce them to standard temperature conditions.*

Attached Thermo-meter (°F)	Barometer Reading (Inches)										
	26·0	26·5	27·0	27·5	28·0	28·5	29·0	29·5	30·0	30·5	31·0
40	·024	·024	·025	·026	·026	·027	·027	·028	·028	·029	·029
41	·026	·027	·028	·028	·029	·029	·030	·030	·031	·031	·032
42	·029	·030	·030	·031	·031	·032	·033	·033	·034	·034	·035
43	·031	·032	·033	·033	·034	·035	·035	·036	·037	·037	·038
44	·034	·035	·035	·036	·037	·037	·038	·039	·039	·040	·041
45	·036	·037	·038	·039	·039	·040	·041	·042	·042	·043	·044
46	·039	·040	·040	·041	·042	·043	·044	·044	·045	·046	·047
47	·041	·042	·043	·044	·045	·046	·046	·047	·048	·049	·050
48	·044	·045	·046	·046	·047	·048	·049	·050	·051	·052	·053
49	·046	·047	·048	·049	·050	·051	·052	·053	·054	·055	·056
50	·049	·050	·051	·052	·053	·054	·055	·056	·057	·058	·058
51	·051	·052	·053	·054	·055	·056	·057	·058	·059	·060	·061
52	·054	·055	·056	·057	·058	·059	·060	·061	·062	·063	·064
53	·056	·057	·058	·060	·061	·062	·063	·064	·065	·066	·067
54	·059	·060	·061	·062	·063	·064	·066	·067	·068	·069	·070
55	·061	·062	·064	·065	·066	·067	·068	·070	·071	·072	·073
56	·064	·065	·066	·067	·069	·070	·071	·072	·074	·075	·076
57	·066	·067	·069	·070	·071	·073	·074	·075	·076	·078	·079
58	·069	·070	·071	·073	·074	·075	·077	·078	·079	·081	·082
59	·071	·072	·074	·075	·077	·078	·079	·081	·082	·083	·085
60	·074	·075	·076	·078	·079	·081	·082	·084	·085	·086	·088
61	·076	·078	·079	·080	·082	·083	·085	·086	·088	·089	·091
62	·079	·080	·082	·083	·085	·086	·088	·089	·091	·092	·094
63	·081	·083	·084	·087	·089	·089	·090	·092	·093	·095	·097
64	·083	·085	·087	·088	·090	·091	·093	·095	·096	·098	·099
65	·086	·088	·089	·091	·093	·094	·096	·097	·099	·101	·102
66	·088	·090	·092	·094	·095	·097	·099	·100	·102	·104	·105
67	·091	·093	·094	·096	·098	·100	·101	·103	·105	·107	·108
68	·093	·095	·097	·099	·101	·102	·104	·106	·108	·109	·111
69	·096	·098	·100	·101	·103	·105	·107	·109	·110	·112	·114
70	·098	·100	·102	·104	·106	·108	·110	·111	·113	·115	·117
71	·101	·103	·105	·107	·108	·110	·112	·114	·116	·118	·120
72	·103	·105	·107	·109	·111	·113	·115	·117	·119	·121	·123
73	·106	·108	·110	·112	·114	·116	·118	·120	·122	·124	·126
74	·108	·110	·112	·114	·116	·118	·121	·123	·125	·127	·129
75	·111	·113	·115	·117	·119	·121	·123	·125	·127	·130	·132
76	·113	·115	·117	·120	·122	·124	·126	·128	·130	·132	·135
77	·116	·118	·120	·122	·124	·127	·129	·131	·133	·135	·137
78	·118	·120	·123	·125	·127	·129	·131	·134	·136	·138	·140
79	·121	·123	·125	·127	·130	·132	·134	·136	·139	·141	·143
80	·123	·125	·128	·130	·132	·135	·137	·139	·142	·144	·146
81	·125	·128	·130	·133	·135	·137	·140	·142	·144	·147	·149
82	·128	·130	·133	·135	·138	·140	·142	·145	·147	·150	·152
83	·130	·133	·135	·138	·140	·143	·145	·148	·150	·153	·155
84	·133	·135	·138	·140	·143	·145	·148	·150	·153	·155	·158
85	·135	·138	·140	·143	·146	·148	·151	·153	·156	·158	·161
86	·138	·140	·143	·146	·148	·151	·153	·156	·159	·161	·164
87	·140	·143	·146	·148	·151	·153	·156	·159	·161	·164	·167
88	·143	·145	·148	·151	·153	·156	·159	·162	·164	·167	·170
89	·145	·148	·151	·153	·156	·159	·162	·164	·167	·170	·172
90	·148	·150	·153	·156	·159	·162	·164	·167	·170	·173	·175
91	·150	·153	·156	·159	·161	·164	·167	·170	·173	·175	·178
92	·153	·155	·158	·161	·164	·167	·170	·173	·175	·178	·181
93	·155	·158	·161	·164	·167	·170	·172	·175	·178	·181	·184
94	·158	·160	·163	·166	·169	·172	·175	·178	·181	·184	·187
95	·160	·163	·166	·169	·172	·175	·178	·181	·184	·187	·190
96	·162	·165	·169	·172	·175	·178	·181	·184	·187	·190	·193
97	·165	·168	·171	·174	·177	·180	·183	·186	·190	·193	·196
98	·167	·170	·174	·177	·180	·183	·186	·189	·192	·196	·199
99	·170	·173	·176	·179	·183	·186	·189	·192	·195	·198	·202
100	·172	·175	·179	·182	·185	·188	·192	·195	·198	·201	·204

TABLE 2
TEMPERATURE CORRECTION OF THE MET. O. KEW-PATTERN BAROMETER MK 2
(Inch Scale)

To be used with barometers having National Physical Laboratory certificates dated ON OR AFTER 1 JANUARY, 1955.

The following corrections are to be **subtracted** *from the barometer readings to reduce them to standard temperature conditions.*

Attached Thermo-meter (°F)	Barometer Reading (Inches)										
	26·0	26·5	27·0	27·5	28·0	28·5	29·0	29·5	30·0	30·5	31·0
40	·020	·020	·021	·021	·021	·022	·022	·022	·023	·023	·024
41	·022	·023	·023	·024	·024	·024	·025	·025	·026	·026	·026
42	·025	·025	·026	·026	·027	·027	·028	·028	·029	·029	·029
43	·027	·028	·028	·029	·029	·030	·030	·031	·031	·032	·032
44	·030	·030	·031	·031	·032	·033	·033	·034	·034	·035	·035
45	·032	·033	·034	·034	·035	·035	·036	·036	·037	·038	·038
46	·035	·035	·036	·037	·037	·038	·039	·039	·040	·041	·041
47	·037	·038	·039	·039	·040	·041	·041	·042	·043	·043	·044
48	·040	·041	·041	·042	·043	·043	·044	·045	·046	·046	·047
49	·042	·043	·044	·045	·045	·046	·047	·048	·048	·049	·050
50	·045	·046	·046	·047	·048	·049	·050	·050	·051	·052	·053
51	·047	·048	·049	·050	·051	·051	·052	·053	·054	·055	·056
52	·050	·051	·051	·052	·053	·054	·055	·056	·057	·058	·059
53	·052	·053	·054	·055	·056	·057	·058	·059	·060	·061	·062
54	·055	·056	·057	·058	·059	·060	·061	·062	·063	·064	·065
55	·057	·058	·059	·060	·061	·062	·063	·064	·065	·066	·068
56	·060	·061	·062	·063	·064	·065	·066	·067	·068	·069	·071
57	·062	·063	·064	·065	·067	·068	·069	·070	·071	·072	·074
58	·064	·066	·067	·068	·069	·070	·072	·073	·074	·075	·077
59	·067	·068	·069	·071	·072	·073	·074	·076	·077	·078	·080
60	·069	·071	·072	·073	·075	·076	·077	·078	·080	·081	·083
61	·072	·073	·075	·076	·077	·078	·080	·081	·082	·084	·086
62	·074	·076	·077	·078	·080	·081	·083	·084	·085	·087	·089
63	·077	·078	·080	·081	·082	·084	·085	·087	·088	·089	·092
64	·079	·081	·082	·084	·085	·087	·088	·090	·091	·092	·094
65	·082	·083	·085	·086	·088	·089	·091	·092	·094	·095	·097
66	·084	·086	·087	·089	·090	·092	·094	·095	·097	·098	·100
67	·087	·088	·090	·091	·093	·095	·096	·098	·099	·101	·103
68	·089	·091	·092	·094	·096	·097	·099	·101	·102	·104	·106
69	·092	·093	·095	·097	·098	·100	·102	·103	·105	·107	·108
70	·094	·096	·098	·099	·101	·103	·104	·106	·108	·110	·111
71	·097	·098	·100	·102	·104	·105	·107	·109	·111	·112	·114
72	·099	·101	·103	·105	·106	·108	·110	·112	·114	·115	·117
73	·102	·103	·105	·107	·109	·111	·113	·115	·116	·118	·120
74	·104	·106	·108	·110	·112	·114	·115	·117	·119	·121	·123
75	·106	·108	·110	·112	·114	·116	·118	·120	·122	·124	·126
76	·109	·111	·113	·115	·117	·119	·121	·123	·125	·127	·129
77	·111	·113	·115	·118	·120	·122	·124	·126	·128	·130	·132
78	·114	·116	·118	·120	·122	·124	·126	·128	·131	·133	·135
79	·116	·118	·121	·123	·125	·127	·129	·131	·133	·135	·138
80	·119	·121	·123	·125	·127	·130	·132	·134	·136	·138	·141
81	·121	·123	·126	·128	·130	·132	·135	·137	·139	·141	·143
82	·124	·126	·128	·131	·133	·135	·137	·140	·142	·144	·146
83	·126	·129	·131	·133	·135	·138	·140	·142	·145	·147	·149
84	·129	·131	·133	·136	·138	·140	·143	·145	·147	·150	·152
85	·131	·134	·136	·138	·141	·143	·145	·148	·150	·153	·155
86	·134	·136	·138	·141	·143	·146	·148	·151	·153	·156	·158
87	·136	·139	·141	·144	·146	·148	·151	·153	·156	·158	·161
88	·139	·141	·144	·146	·149	·151	·154	·156	·159	·161	·164
89	·141	·144	·146	·149	·151	·154	·156	·159	·162	·164	·167
90	·143	·146	·149	·151	·154	·157	·159	·162	·164	·167	·170

TABLE 3

CORRECTION OF THE MET. O. KEW-PATTERN BAROMETER MK 1 (INCH SCALE) TO STANDARD GRAVITY IN LATITUDE 45°

To be used with barometers having National Physical Laboratory certificate dated ON OR BEFORE 31 *DECEMBER* 1954.

These corrections are to be subtracted *for latitudes* 0°–44° *and* added *for latitudes* 46°–90°.

Lat. N or S (subtract correction)	Correction At 29 in.	Correction At 31 in.	Lat. N or S (add correction)	Lat. N or S (subtract correction)	Correction At 29 in.	Correction At 31 in.	Lat. N or S (add correction)
°	in.	in.	°	°	in.	in.	°
22	·055	·058	68	45	·000	·000	45
21	·056	·060	69	44	·002	·002	46
20	·058	·062	70	43	·005	·005	47
19	·060	·064	71	42	·008	·008	48
18	·062	·066	72	41	·010	·011	49
17	·063	·067	73	40	·013	·014	50
16	·065	·069	74	39	·015	·016	51
15	·066	·070	75	38	·018	·019	52
14	·067	·072	76	37	·021	·022	53
13	·068	·073	77	36	·023	·025	54
12	·070	·074	78	35	·026	·027	55
11	·071	·075	79	34	·028	·030	56
10	·072	·076	80	33	·031	·033	57
9	·072	·077	81	32	·033	·035	58
8	·073	·078	82	31	·036	·038	59
7	·074	·079	83	30	·038	·040	60
6	·075	·080	84	29	·040	·043	61
5	·075	·080	85	28	·042	·045	62
4	·075	·080	86	27	·045	·048	63
3	·075	·080	87	26	·047	·050	64
2	·075	·080	88	25	·049	·052	65
1	·075	·080	89	24	·051	·054	66
0	·076	·081	90	23	·053	·056	67

TABLE 4

CORRECTION OF THE MET. O. KEW-PATTERN BAROMETER MK 2 (INCH SCALE) TO STANDARD GRAVITY, i.e. 9.80665 m/s²

To be used with barometers having National Physical Laboratory certificate dated ON OR AFTER 1 *JANUARY* 1955.

Lat. N or S	Correction At 29 in.	Correction At 31 in.	Lat. N or S	Correction At 29 in.	Correction At 31 in.	Lat. N or S	Correction At 29 in.	Correction At 31 in.	Lat. N or S	Correction At 29 in.	Correction At 31 in.
°	in.	in.	°	in	in.	°	in.	in.	°	in.	in.
0	−·078	−·083	25	−·051	−·054	46	+·001	+·001	67	+·052	+·055
			26	·048	·052	47	·004	·004	68	·054	·057
5	−·077	−·082	27	·046	·050	48	·007	·007	69	·055	·059
6	·076	·081	28	·044	·047	49	·009	·010	70	·057	·061
7	·075	·081	29	·042	·045	50	·012	·013	71	·059	·063
8	·075	·080	30	·040	·042	51	·014	·015	72	·061	·065
9	·074	·079	31	·037	·040	52	·017	·018	73	·062	·066
10	·073	·078	32	·035	·037	53	·020	·021	74	·064	·068
11	·072	·077	33	·033	·035	54	·022	·024	75	·065	·069
12	·071	·076	34	·030	·032	55	·025	·026	76	·066	·071
13	·070	·075	35	·028	·029	56	·027	·029	77	·067	·072
14	·069	·074	36	·025	·027	57	·030	·032	78	·069	·073
15	·068	·072	37	·023	·024	58	·032	·034	79	·070	·074
16	·066	·071	38	·020	·021	59	·035	·037	80	·071	·075
17	·065	·069	39	·017	·019	60	·037	·039	81	·071	·076
18	·063	·068	40	·015	·016	61	·039	·042	82	·072	·077
19	·062	·066	41	·012	·013	62	·041	·044	83	·073	·078
20	·060	·064	42	·009	·010	63	·044	·047	84	·074	·079
21	·058	·062	43	·007	·007	64	·046	·049	85	+·074	+·079
22	·056	·060	44	·004	·004	65	·048	·051			
23	·055	·058	45	−·001	−·002	66	+·050	+·053	90	+·075	+·080
24	−·053	−·056									

TABLE 5

CORRECTION OF INCH BAROMETERS TO MEAN SEA LEVEL

*These corrections are to be **added** to the barometer readings.*

Height in feet	Temperature of Air (Dry Bulb in Screen), °F										Height in feet
	0°	10°	20°	30°	40°	50°	60°	70°	80°	90°	
5	·006	·006	·006	·006	·006	·006	·006	·005	·005	·005	5
10	·012	·012	·012	·011	·011	·011	·011	·010	·010	·010	10
15	·019	·018	·018	·017	·017	·017	·017	·016	·016	·015	15
20	·025	·024	·023	·023	·023	·022	·022	·021	·021	·020	20
25	·031	·030	·029	·029	·029	·028	·027	·027	·026	·026	25
30	·037	·036	·035	·035	·034	·033	·032	·032	·031	·031	30
35	·043	·042	·041	·041	·040	·039	·038	·037	·037	·036	35
40	·049	·048	·047	·046	·045	·044	·043	·042	·042	·041	40
45	·056	·054	·053	·052	·051	·050	·049	·048	·047	·046	45
50	·062	·060	·059	·058	·056	·055	·054	·053	·052	·051	50
55	·068	·066	·065	·064	·062	·061	·060	·059	·057	·056	55
60	·074	·072	·071	·069	·068	·066	·065	·064	·062	·061	60
65	·080	·078	·077	·075	·074	·072	·071	·069	·068	·066	65
70	·086	·084	·083	·081	·079	·077	·076	·074	·073	·071	70
75	·092	·090	·089	·087	·085	·083	·082	·080	·078	·076	75
80	·098	·096	·094	·092	·091	·089	·087	·085	·083	·081	80
85	·105	·102	·100	·098	·097	·095	·093	·090	·089	·087	85
90	·111	·108	·106	·104	·102	·101	·098	·095	·094	·092	90
95	·117	·114	·112	·110	·108	·106	·103	·101	·099	·097	95
100	·123	·120	·118	·115	·113	·111	·108	·106	·104	·101	100

TABLE 6

TEMPERATURE CORRECTION OF THE MET. O. KEW-PATTERN BAROMETER MK 1
(Millibar Scale)

To be used with barometers having National Physical Laboratory certificate dated ON OR BEFORE 31 DECEMBER, 1954.
*These corrections are to be **subtracted** from the barometer readings when the attached thermometer is ABOVE 285° A, and **added** when it is BELOW 285° A, to reduce the barometer readings to 285° A.*

Attached thermometer (add correction)	Barometer readings (mb)										Attached thermometer (subtract correction)
	880	900	920	940	960	980	1000	1020	1040	1060	
284°A	0·1	0·1	0·2	0·2	0·2	0·2	0·2	0·2	0·2	0·2	286°A
283	0·3	0·3	0·3	0·3	0·3	0·3	0·3	0·3	0·4	0·4	287
282	0·5	0·5	0·5	0·5	0·5	0·5	0·5	0·5	0·5	0·5	288
281	0·6	0·6	0·6	0·6	0·7	0·7	0·7	0·7	0·7	0·7	289
280	0·8	0·8	0·8	0·8	0·8	0·8	0·9	0·9	0·9	0·9	290
279	0·9	0·9	0·9	1·0	1·0	1·0	1·0	1·1	1·1	1·1	291
278	1·1	1·1	1·1	1·1	1·1	1·2	1·2	1·2	1·2	1·3	292
277	1·2	1·2	1·3	1·3	1·3	1·3	1·4	1·4	1·4	1·5	292
276	1·4	1·4	1·4	1·5	1·5	1·5	1·5	1·6	1·6	1·6	294
275	1·5	1·5	1·6	1·6	1·6	1·7	1·7	1·7	1·8	1·8	295
274	1·7	1·7	1·7	1·8	1·8	1·9	1·9	1·9	1·9	2·0	296
273	1·8	1·9	1·9	1·9	2·0	2·0	2·1	2·1	2·1	2·2	297
272	2·0	2·0	2·1	2·1	2·1	2·2	2·2	2·3	2·3	2·3	298
271	2·1	2·2	2·2	2·3	2·3	2·3	2·4	2·4	2·5	2·5	299
270	2·3	2·3	2·4	2·4	2·5	2·5	2·6	2·6	2·7	2·7	300
269	2·4	2·5	2·5	2·6	2·6	2·7	2·7	2·8	2·8	2·9	301
268	2·6	2·6	2·7	2·7	2·8	2·9	2·9	3·0	3·0	3·1	302
267	2·7	2·8	2·8	2·9	3·0	3·0	3·1	3·1	3·2	3·3	303
266	2·9	2·9	3·0	3·1	3·1	3·2	3·3	3·3	3·4	3·4	304
265	3·0	3·1	3·2	3·2	3·3	3·3	3·4	3·5	3·5	3·6	305
264	3·2	3·3	3·3	3·4	3·5	3·5	3·6	3·7	3·7	3·8	306
263	3·3	3·4	3·5	3·5	3·6	3·7	3·7	3·8	3·9	4·0	307
262	3·5	3·6	3·6	3·7	3·8	3·9	3·9	4·0	4·1	4·2	308
261	3·6	3·7	3·8	3·9	3·9	4·0	4·1	4·2	4·3	4·3	309
260	3·8	3·9	3·9	4·0	4·1	4·2	4·3	4·4	4·4	4·5	310
259	3·9	4·0	4·1	4·2	4·3	4·4	4·5	4·5	4·6	4·7	311
258	4·1	4·2	4·3	4·3	4·4	4·5	4·6	4·7	4·8	4·9	312
257	4·2	4·3	4·4	4·5	4·6	4·7	4·9	4·9	5·0	5·1	313
256	4·4	4·5	4·6	4·7	4·8	4·9	5·0	5·1	5·1	5·2	314
255	4·5	4·6	4·7	4·8	4·9	5·0	5·1	5·2	5·3	5·4	315

TABLE 7

TEMPERATURE CORRECTION OF THE MET. O. KEW-PATTERN BAROMETER MK 2
(Millibar Scale)

To be used with barometers having National Physical Laboratory certificate dated ON OR AFTER 1 JANUARY, 1955.

These corrections are to be **subtracted** *from the barometer readings to reduce them to 0° C.*

Attached Thermometer		Barometer Readings (Millibars)						
°A	°C	920	940	960	980	1000	1020	1040
273	0	0·00	0·00	0·00	0·00	0·00	0·00	0·00
274	1	0·16	0·16	0·16	0·17	0·17	0·17	0·18
275	2	0·32	0·32	0·33	0·34	0·34	0·35	0·36
276	3	0·48	0·48	0·49	0·50	0·51	0·52	0·53
277	4	0·63	0·64	0·65	0·67	0·68	0·70	0·71
278	5	0·79	0·80	0·82	0·84	0·85	0·87	0·89
279	6	0·95	0·97	0·99	1·01	1·03	1·05	1·07
280	7	1·11	1·13	1·15	1·18	1·20	1·22	1·25
281	8	1·26	1·29	1·32	1·35	1·38	1·40	1·43
282	9	1·42	1·45	1·48	1·52	1·55	1·57	1·60
283	10	1·58	1·61	1·64	1·68	1·71	1·74	1·77
284	11	1·74	1·77	1·81	1·84	1·88	1·91	1·95
285	12	1·89	1·93	1·97	2·01	2·05	2·09	2·13
286	13	2·05	2·09	2·13	2·18	2·22	2·26	2·30
287	14	2·21	2·25	2·30	2·34	2·39	2·44	2·48
288	15	2·36	2·41	2·46	2·51	2·56	2·61	2·66
289	16	2·52	2·57	2·63	2·68	2·73	2·78	2·83
290	17	2·68	2·73	2·79	2·84	2·90	2·96	3·01
291	18	2·84	2·89	2·95	3·01	3·07	3·13	3·19
292	19	2·99	3·05	3·12	3·18	3·24	3·30	3·36
293	20	3·15	3·22	3·28	3·35	3·41	3·48	3·54
294	21	3·31	3·38	3·44	3·51	3·58	3·65	3·72
295	22	3·46	3·54	3·61	3·68	3·75	3·82	3·89
296	23	3·62	3·70	3·77	3·84	3·92	3·99	4·07
297	24	3·78	3·86	3·93	4·01	4·09	4·17	4·25
298	25	3·93	4·02	4·10	4·18	4·26	4·34	4·42
299	26	4·09	4·18	4·26	4·34	4·43	4·51	4·60
300	27	4·25	4·34	4·42	4·51	4·60	4·69	4·77
301	28	4·40	4·49	4·59	4·68	4·77	4·86	4·95
302	29	4·56	4·65	4·75	4·84	4·94	5·03	5·13
303	30	4·72	4·81	4·91	5·01	5·11	5·20	5·30
304	31	4·88	4·97	5·07	5·18	5·28	5·37	5·47
305	32	5·03	5·13	5·24	5·34	5·44	5·55	5·65
306	33	5·19	5·29	5·40	5·51	5·61	5·72	5·82
307	34	5·34	5·45	5·56	5·67	5·78	5·89	6·00
308	35	5·50	5·61	5·72	5·84	5·95	6·06	6·17
309	36	5·65	5·76	5·88	6·00	6·11	6·23	6·35
310	37	5·81	5·92	6·04	6·17	6·28	6·40	6·53
311	38	5·97	6·09	6·21	6·34	6·46	6·58	6·71
312	39	6·13	6·25	6·37	6·51	6·63	6·75	6·88
313	40	6·28	6·41	6·54	6·67	6·80	6·93	7·06

TABLE 8

CORRECTION OF THE MET. O. KEW-PATTERN BAROMETER MK 1 (MILLIBAR SCALE) TO STANDARD GRAVITY IN LATITUDE 45°

To be used with barometers having National Physical Laboratory certificate dated ON OR BEFORE 31 DECEMBER, 1954.

These corrections are to be **subtracted** *for latitudes 0°–44° and* **added** *for latitudes 46°–90°.*

Lat. N or S (subtract correction)	Correction At 980 mb	Correction At 1040 mb	Lat. N or S (add correction)	Lat. N or S (subtract correction)	Correction At 980 mb	Correction At 1040 mb	Lat. N or S (add correction)
°	mb.	mb.	°	°	mb.	mb.	°
24	1·73	1·83	66	45	0·00	0·00	45
23	1·79	1·91	67	44	0·09	0·09	46
22	1·86	1·97	68	43	0·18	0·19	47
21	1·92	2·04	69	42	0·27	0·29	48
20	1·98	2·10	70	41	0·36	0·38	49
19	2·03	2·16	71	40	0·45	0·47	50
18	2·09	2·22	72	39	0·54	0·57	51
17	2·14	2·28	73	38	0·63	0·67	52
16	2·19	2·32	74	37	0·71	0·75	53
15	2·23	2·38	75	36	0·80	0·85	54
14	2·29	2·42	76	35	0·89	0·94	55
13	2·32	2·46	77	34	0·97	1·03	56
12	2·36	2·50	78	33	1·05	1·11	57
11	2·39	2·54	79	32	1·13	1·20	58
10	2·42	2·57	80	31	1·21	1·29	59
9	2·46	2·61	81	30	1·29	1·37	60
8	2·48	2·63	82	29	1·37	1·45	61
7	2·50	2·66	83	28	1·44	1·53	62
6	2·53	2·68	84	27	1·52	1·61	63
5	2·54	2·70	85	26	1·59	1·69	64
0	2·59	2·75	90	25	1·66	1·76	65

TABLE 9

CORRECTION OF THE MET. O. KEW-PATTERN BAROMETER MK 2 (MILLIBAR SCALE) TO STANDARD GRAVITY, i.e. 9.80665 m/s^2

To be used with barometers having National Physical Laboratory certificate dated ON OR AFTER 1 JANUARY, 1955.

Lat. N or S	Correction At 980 mb	Correction At 1040 mb	Lat. N or S	Correction At 980 mb	Correction At 1040 mb	Lat. N or S	Correction At 980 mb	Correction At 1040 mb	Lat. N or S	Correction At 980 mb	Correction At 1040 mb
°	mb.	mb.	°	mb.	mb.	°	mb.	mb.	°	mb.	mb.
0	−2·63	−2·79	25	−1·71	−1·81	46	+0·04	+0·04	67	+1·75	+1·86
			26	1·64	1·74	47	0·13	0·14	68	1·81	1·92
5	−2·59	−2·75	27	1·57	1·66	48	0·22	0·23	69	1·87	1·99
6	2·57	2·73	28	1·49	1·58	49	0·31	0·33	70	1·93	2·05
7	2·55	2·71	29	1·42	1·50	50	0·40	0·42	71	1·99	2·11
8	2·53	2·68	30	1·34	1·42	51	0·49	0·52	72	2·05	2·17
9	2·50	2·66	31	1·26	1·34	52	0·58	0·61	73	2·10	2·23
10	2·47	2·62	32	1·18	1·25	53	0·66	0·70	74	2·15	2·28
11	2·44	2·59	33	1·10	1·17	54	0·75	0·80	75	2·19	2·33
12	2·41	2·55	34	1·02	1·08	55	0·84	0·89	76	2·24	2·37
13	2·37	2·51	35	0·93	0·99	56	0·92	0·98	77	2·28	2·42
14	2·33	2·47	36	0·85	0·90	57	1·00	1·06	78	2·32	2·46
15	2·28	2·42	37	0·76	0·81	58	1·09	1·15	79	2·35	2·50
16	2·24	2·37	38	0·67	0·72	59	1·17	1·24	80	2·38	2·53
17	2·19	2·32	39	0·59	0·62	60	1·24	1·32	81	2·41	2·56
18	2·14	2·27	40	0·50	0·53	61	1·32	1·40	82	2·44	2·59
19	2·08	2·21	41	0·41	0·43	62	1·40	1·48	83	2·46	2·62
20	2·03	2·15	42	0·32	0·34	63	1·47	1·56	84	2·48	2·64
21	1·97	2·09	43	0·23	0·24	64	1·54	1·64	85	+2·50	+2·66
22	1·91	2·02	44	0·14	0·15	65	1·61	1·71			
23	1·84	1·95	45	−0·05	−0·05	66	+1·68	+1·79	90	+2·54	+2·70
24	−1·78	−1·88									

TABLE 10

CORRECTION OF MILLIBAR BAROMETERS TO MEAN SEA LEVEL

These corrections are to be **added** *to the barometer readings.*

Height	Temperature of Air (°C) (Dry bulb in screen)											
	−15	−10	−5	0	5	10	15	20	25	30	35	40
metres					*millibars*							
5	0·7	0·6	0·6	0·6	0·6	0·6	0·6	0·6	0·6	0·6	0·6	0·5
10	1·3	1·3	1·3	1·3	1·2	1·2	1·2	1·2	1·1	1·1	1·1	1·1
15	2·0	1·9	1·9	1·9	1·8	1·8	1·8	1·7	1·7	1·7	1·7	1·6
20	2·6	2·6	2·5	2·5	2·5	2·4	2·4	2·3	2·3	2·3	2·2	2·2
25	3·3	3·2	3·2	3·1	3·1	3·0	3·0	2·9	2·9	2·8	2·8	2·7
30	4·0	3·9	3·8	3·8	3·7	3·6	3·6	3·5	3·4	3·4	3·3	3·3
35	4·6	4·5	4·5	4·4	4·3	4·2	4·2	4·1	4·0	3·9	3·9	3·8
40	5·3	5·2	5·1	5·0	4·9	4·8	4·7	4·7	4·6	4·5	4·4	4·4
45	6·0	5·9	5·7	5·6	5·5	5·4	5·3	5·3	5·2	5·1	5·0	4·9
50	6·6	6·5	6·4	6·3	6·2	6·0	5·9	5·8	5·7	5·6	5·6	5·5

TABLE 11

APPROXIMATE BAROMETER CORRECTIONS FOR TEMPERATURE AND HEIGHT

	Inches	Millibars*
Temperature ...	Subtract 0·003 inch for each degree F the attached thermometer reads above freezing point.	Subtract the temperature of the attached thermometer, in degrees A, from 285°A (taking account of sign), and divide by 6, to get the correction in millibars.
Height	The height in feet, increased by 10 per cent, gives the correction in thousandths of an inch.	The height in feet, increased by 10 per cent, and divided by 30, gives the correction in millibars.

Examples:

Inch barometer	attached thermometer 56°F	Temperature correction	− ·072 in
	height 90 feet	Height ,,	+ ·099 ,,
Millibar barometer	(NPL certificate before 1 January, 1955):		
	attached thermometer 299°A	Temperature ,,	− 2·3 mb
	height 90 feet	Height ,,	+ 3·3 ,,

*For barometers with NPL certificate dated on or after 1 January, 1955, the temperature of the attached thermometer should be subtracted from 273°A or, if the thermometer is graduated in °C, from 0°C.

THE DIURNAL VARIATION OF BAROMETRIC PRESSURE IN THE ZONES OF LATITUDE 0°–10° AND 10°–20°, N OR S

TABLE 12

CORRECTION TO BE APPLIED TO THE OBSERVED PRESSURE FOR DIURNAL VARIATION

Local Time	0°–10° N or S		10°–20° N or S	
	mb	inches	mb	inches
0	−0·6	− ·018	−0·5	− ·015
1	−0·1	− ·003	−0·1	− ·003
2	+0·3	+ ·009	+0·3	+ ·009
3	+0·7	+ ·021	+0·7	+ ·021
4	+0·8	+ ·024	+0·8	+ ·024
5	+0·6	+ ·018	+0·6	+ ·018
6	+0·2	+ ·006	+0·2	+ ·006
7	−0·4	− ·012	−0·3	− ·009
8	−0·9	− ·027	−0·8	− ·024
9	−1·3	− ·038	−1·1	− ·032
10	−1·4	− ·041	−1·2	− ·035
11	−1·1	− ·032	−1·0	− ·030
12	−0·6	− ·018	−0·5	− ·015
13	+0·1	+ ·003	+0·1	+ ·003
14	+0·7	+ ·021	+0·7	+ ·021
15	+1·3	+ ·038	+1·1	+ ·032
16	+1·5	+ ·044	+1·3	+ ·038
17	+1·4	+ ·041	+1·2	+ ·035
18	+1·0	+ ·030	+0·9	+ ·027
19	+0·5	+ ·015	+0·3	+ ·009
20	−0·1	− ·003	−0·2	− ·006
21	−0·6	− ·018	−0·6	− ·018
22	−0·9	− ·027	−0·8	− ·024
23	−0·9	− ·027	−0·8	− ·024
24	−0·6	− ·018	−0·5	− ·015

TABLE 13

AVERAGE VALUES OF THE BAROMETRIC CHANGE IN AN HOUR, DUE TO THE DIURNAL VARIATION

Local Time	0°–10° N or S		10°–20° N or S	
	mb	inches	mb	inches
0–1	−0·5	− ·015	−0·4	− ·012
1–2	−0·4	− ·012	−0·4	− ·012
2–3	−0·4	− ·012	−0·4	− ·012
3–4	−0·1	− ·003	−0·1	− ·003
4–5	+0·2	+ ·006	+0·2	+ ·006
5–6	+0·4	+ ·012	+0·4	+ ·012
6–7	+0·6	+ ·018	+0·5	+ ·015
7–8	+0·5	+ ·015	+0·5	+ ·015
8–9	+0·4	+ ·012	+0·3	+ ·009
9–10	+0·1	+ ·003	+0·1	+ ·003
10–11	−0·3	− ·009	−0·2	− ·006
11–12	−0·5	− ·015	−0·5	− ·015
12–13	−0·7	− ·021	−0·6	− ·018
13–14	−0·6	− ·018	−0·6	− ·018
14–15	−0·6	− ·018	−0·4	− ·012
15–16	−0·2	− ·006	−0·2	− ·006
16–17	+0·1	+ ·003	+0·1	+ ·003
17–18	+0·4	+ ·012	+0·3	+ ·009
18–19	+0·5	+ ·015	+0·6	+ ·018
19–20	+0·6	+ ·018	+0·5	+ ·015
20–21	+0·5	+ ·015	+0·4	+ ·012
21–22	+0·3	+ ·009	+0·2	+ ·006
22–23	0	·000	0	·000
23–24	−0·3	− ·009	−0·3	− ·009

These tables are based on observations made in British ships, at the hours 0000, 0400, 0800, 1200, 1600, 2000 local time, between 1919–38.

In the tropics, should the barometer, after correction for diurnal variation (Table 12), be as much as 3 millibars (approximately 0·1 inch) below the monthly normal for the locality, as shown on meteorological charts, the mariner should be on the alert, as there is a distinct possibility that a tropical storm has formed, or is forming. A comparison of subsequent hourly changes in his barometer with the corresponding figures in Table 13 will show whether these changes indicate a real further fall in pressure and, if so, its amount.

Caution: When entering a barometric pressure in the log, or when including it in a wireless weather report, the correction for diurnal variation must not be applied.

TABLE 14

EQUIVALENTS IN MILLIBARS OF INCHES OF MERCURY AT 0°C (32°F) AND STANDARD GRAVITY 9.80665 m/s²

(Hundredths of an inch)

Inches	0	1	2	3	4	5	6	7	8	9
						Millibars				
27·0	914·3	914·7	915·0	915·3	915·7	916·0	916·4	916·7	917·0	917·4
27·1	917·7	918·1	918·4	918·7	919·1	919·4	919·7	920·1	920·4	920·8
27·2	921·1	921·4	921·8	922·1	922·5	922·8	923·1	923·5	923·8	924·1
27·3	924·5	924·8	925·2	925·5	925·8	926·2	926·5	926·9	927·2	927·5
27·4	927·9	928·2	928·5	928·9	929·2	929·6	929·9	930·2	930·6	930·9
27·5	931·3	931·6	931·9	932·3	932·6	933·0	933·3	933·6	934·0	934·3
27·6	934·6	935·0	935·3	935·7	936·0	936·3	936·7	937·0	937·4	937·7
27·7	938·0	938·4	938·7	939·0	939·4	939·7	940·1	940·4	940·7	941·1
27·8	941·4	941·8	942·1	942·4	942·8	943·1	943·4	943·8	944·1	944·5
27·9	944·8	945·1	945·5	945·8	946·2	946·5	946·8	947·2	947·5	947·9
28·0	948·2	948·5	948·9	949·2	949·5	949·9	950·2	950·6	950·9	951·2
28·1	951·6	951·9	952·3	952·6	952·9	953·3	953·6	953·9	954·3	954·6
28·2	955·0	955·3	955·6	956·0	956·3	956·7	957·0	957·3	957·7	958·0
28·3	958·3	958·7	959·0	959·4	959·7	960·0	960·4	960·7	961·1	961·4
28·4	961·7	962·1	962·4	962·8	963·1	963·4	963·8	964·1	964·4	964·8
28·5	965·1	965·5	965·8	966·1	966·5	966·8	967·2	967·5	967·8	968·2
28·6	968·5	968·8	969·2	969·5	969·9	970·2	970·5	970·9	971·2	971·6
28·7	971·9	972·2	972·6	972·9	973·2	973·6	973·9	974·3	974·6	974·9
28·8	975·3	975·6	976·0	976·3	976·6	977·0	977·3	977·7	978·0	978·3
28·9	978·7	979·0	979·3	979·7	980·0	980·4	980·7	981·0	981·4	981·7
29·0	982·1	982·4	982·7	983·1	983·4	983·7	984·1	984·4	984·8	985·1
29·1	985·4	985·8	986·1	986·5	986·8	987·1	987·5	987·8	988·1	988·5
29·2	988·8	989·2	989·5	989·8	990·2	990·5	990·9	991·2	991·5	991·9
29·3	992·2	992·6	992·9	993·2	993·6	993·9	994·2	994·6	994·9	995·3
29·4	995·6	995·9	996·3	996·6	997·0	997·3	997·6	998·0	998·3	998·6

TABLE 14—(contd)

Inches	0	1	2	3	4	5	6	7	8	9
					Millibars					
29·5	999·0	999·3	999·7	1000·0	1000·3	1000·7	1001·0	1001·4	1001·7	1002·0
29·6	1002·4	1002·7	1003·0	1003·4	1003·7	1004·1	1004·4	1004·7	1005·1	1005·4
29·7	1005·8	1006·1	1006·4	1006·8	1007·1	1007·5	1007·8	1008·1	1008·5	1008·8
29·8	1009·1	1009·5	1009·8	1010·2	1010·5	1010·8	1011·2	1011·5	1011·9	1012·2
29·9	1012·5	1012·9	1013·2	1013·5	1013·9	1014·2	1014·6	1014·9	1015·2	1015·6
30·0	1015·9	1016·3	1016·6	1016·9	1017·3	1017·6	1017·9	1018·3	1018·6	1019·0
30·1	1019·3	1019·6	1020·0	1020·3	1020·7	1021·0	1021·3	1021·7	1022·0	1022·4
30·2	1022·7	1023·0	1023·4	1023·7	1024·0	1024·4	1024·7	1025·1	1025·4	1025·7
30·3	1026·1	1026·4	1026·8	1027·1	1027·4	1027·8	1028·1	1028·4	1028·8	1029·1
30·4	1029·5	1029·8	1030·1	1030·5	1030·8	1031·2	1031·5	1031·8	1032·2	1032·5
30·5	1032·8	1033·2	1033·5	1033·9	1034·2	1034·5	1034·9	1035·2	1035·6	1035·9
30·6	1036·2	1036·6	1036·9	1037·3	1037·6	1037·9	1038·3	1038·6	1038·9	1039·3
30·7	1039·6	1040·0	1040·3	1040·6	1041·0	1041·3	1041·7	1042·0	1042·3	1042·7
30·8	1043·0	1043·3	1043·7	1044·0	1044·4	1044·7	1045·0	1045·4	1045·7	1046·1
30·9	1046·4	1046·7	1047·1	1047·4	1047·7	1048·1	1048·4	1048·8	1049·1	1049·4

To reduce millimetres of pressure to millibars, increase the number of millimetres by one third.

Example   764·8 millimetres

add   254·9

1019·7 millibars

	Thousandths of an Inch		
Inch	Millibar	Inch	Millibar
·001	·0	·006	·2
·002	·1	·007	·2
·003	·1	·008	·3
·004	·1	·009	·3
·005	·2		

o

137

TABLE 15

DEW-POINT (°C)

(FOR USE WITH MARINE SCREEN)

Dry Bulb °C		Depression of Wet Bulb																								Dry Bulb °C
	0°	0·2°	0·4°	0·6°	0·8°	1·0°	1·2°	1·4°	1·6°	1·8°	2·0°	2·5°	3·0°	3·5°	4·0°	4·5°	5·0°	5·5°	6·0°	6·5°	7·0°	7·5°	8·0°	8·5°	9·0°	
40	40	40	40	39	39	39	39	38	38	38	38	37	36	36	35	34	34	33	32	32	31	30	29	29	28	40
39	39	39	39	38	38	38	38	37	37	37	37	36	35	35	34	33	33	32	31	31	30	29	28	28	27	39
38	38	38	38	37	38	37	37	37	36	36	36	35	34	34	33	32	32	31	30	30	29	28	26	26	26	38
37	37	37	37	36	36	36	36	35	35	35	34	34	33	32	32	31	30	30	29	28	28	27	26	25	24	37
36	36	36	36	36	35	35	35	34	34	34	33	33	32	31	31	30	29	29	28	27	26	26	24	23	23	36
35	35	35	34	34	34	34	34	33	33	33	32	32	31	30	30	29	28	28	27	26	25	24	24	23	22	35
34	34	34	33	33	33	33	33	32	32	32	31	31	30	29	29	28	27	26	26	25	24	23	23	22	21	34
33	33	33	32	32	32	32	32	31	31	31	30	30	29	28	28	27	26	25	25	24	23	22	21	20	19	33
32	32	32	31	31	31	31	31	30	30	30	29	29	28	27	26	26	25	24	23	23	22	21	20	19	18	32
31	31	31	30	30	30	30	30	29	29	29	28	28	27	26	25	25	24	23	22	21	20	20	19	18	17	31
30	30	30	29	29	29	29	28	28	28	28	27	27	26	25	24	24	23	22	21	20	19	18	17	17	16	30
29	29	29	28	28	28	28	28	27	27	27	26	25	25	24	23	22	22	21	20	19	18	17	16	15	14	29
28	28	28	27	27	27	27	27	26	26	25	25	24	24	23	22	21	21	20	19	18	17	16	15	14	13	28
27	27	27	27	26	26	26	25	25	25	24	24	23	23	22	21	20	20	19	18	17	16	15	14	13	11	27
26	26	26	25	25	24	24	24	23	23	23	23	22	22	21	20	19	18	17	16	15	14	13	12	11	10	26
25	25	25	24	24	24	23	23	23	22	22	22	21	20	20	19	18	17	16	15	14	13	12	11	10	8	25
24	24	24	23	23	23	22	22	22	21	21	21	20	19	19	18	17	16	15	14	13	12	11	9	8	7	24
23	23	23	22	22	21	21	21	20	20	20	20	19	18	17	17	16	15	14	13	12	11	9	8	7	5	23
22	22	22	21	21	21	20	20	20	19	19	19	18	17	16	15	14	13	12	11	10	9	8	6	5	3	22
21	21	21	20	20	20	19	19	19	18	18	18	17	16	15	14	13	12	11	10	9	8	6	5	3	1	21
20	20	20	19	19	19	18	18	18	17	17	17	16	15	14	13	12	11	10	9	7	6	5	3	1	0	20
19	19	19	18	18	18	17	17	17	16	16	16	15	14	13	12	11	10	9	7	6	4	3	1	0	−2	19
18	18	18	17	17	17	16	16	16	15	15	15	14	13	12	11	10	8	7	6	4	3	1	0	−2	−5	18
17	17	17	16	16	16	15	15	15	14	14	14	13	12	11	9	8	7	6	4	3	1	0	−3	−5	−10	17
16	16	16	15	15	15	14	14	14	13	13	13	11	10	9	8	7	6	4	3	1	0	−2	−5	−7	−14	16
15	15	15	14	14	13	13	13	12	12	12	12	10	9	8	7	6	4	3	1	0	−2	−5	−7	−10	−18	15
14	14	14	13	13	13	12	12	11	11	11	11	9	8	7	6	4	3	1	0	−2	−4	−7	−10	−13	−23	14
13	13	13	12	12	11	11	11	10	10	10	10	8	7	6	4	3	1	0	−2	−4	−7	−10	−13	−17	−33	13
12	12	12	11	11	11	10	10	10	9	9	9	7	6	4	3	1	0	−2	−4	−6	−9	−12	−16	−22		12
11	11	11	11	10	10	10	10	9	8	8	8	6	4	3	1	0	−2	−4	−6	−8	−12	−15	−21	−30		11
10	10	10	9	9	8	8	8	7	7	6	6	4	3	2	0	−2	−3	−6	−8	−11	−15	−19	−27			10

138

TABLE 15—(contd)

Depression of Wet Bulb

Dry Bulb °C	0°	0·2°	0·4°	0·6°	0·8°	1·0°	1·2°	1·4°	1·6°	1·8°	2·0°	2·5°	3·0°	3·5°	4·0°	4·5°	5·0°	5·5°	6·0°	6·5°	7·0°	Dry Bulb °C
9	9	9	8	8	7	7	6	6	5	5	4	3	2	0	−1	−3	−5	−8	−10	−14	−18	9
8	8	8	7	7	6	6	5	5	4	4	3	2	1	−1	−3	−5	−7	−10	−13	−17		8
7	7	7	6	6	5	5	4	4	3	3	2	1	−1	−3	−4	−7	−9	−12	−16			7
6	6	6	6	5	4	4	3	3	2	2	1	−1	−2	−4	−6	−9	−11	−15				6
5	5	5	5	4	3	3	2	2	1	1	0	−2	−4	−6	−8	−11	−14	−18				5
4	4	4	4	3	2	2	1	1	0	0	−1	−3	−5	−7	−10	−14						4
3	3	3	3	2	1	1	0	0	−2	−2	−3	−5	−7	−8	−11	−14	−17					3
2	2	2	2	1	0	0	−1	−2	−3	−3	−4	−5	−8	−10	−13	−16	−19					2
1	1	1	1	0	−1	−1	−2	−3	−4	−4	−5	−7	−9	−12	−15	−19						1
0	0	−1	−1	−2	−2	−3	−4	−4	−5	−6	−7	−9	−11	−14	−18							0
−1	−1	−2	−2	−3	−4	−4	−5	−6	−6	−7	−8	−10	−13	−17								
−2	−2	−3	−4	−4	−5	−6	−7	−7	−8	−9	−10	−12	−15	−19								
−3	−3	−4	−5	−5	−7	−8	−8	−9	−9	−10	−11	−14	−18									
−4	−4	−5	−6	−7	−8	−8	−9	−10	−11	−11	−13	−16										
−5	−5	−6	−7	−8	−9	−10	−10	−11	−13	−14	−15	−18										
−6	−6	−7	−8	−9	−10	−11	−12	−13	−14	−15	−17											
−7	−7	−8	−9	−10	−11	−12	−13	−14	−16	−17	−19											
−8	−8	−9	−10	−11	−13	−14	−15	−16	−18	−19												
−9	−9	−10	−11	−13	−14	−15	−16	−18	−19													
−10	−10	−11	−12	−14	−15	−17	−18															
−11	−11	−12	−13	−15	−16	−17	−18															
−12	−12	−13	−14	−16	−17	−18																
−13	−13	−14	−15	−17	−18																	
−14	−14	−15	−16	−18																		
−15	−15	−16	−17																			
−16	−16	−17	−18																			
−17	−17	−18	−19																			
−18	−18	−19																				
−19	−19																					

In the tables, lines are ruled to draw attention to the fact that above the line evaporation is going on from a water surface, while below the line it is going on from an ice surface. Owing to this interpolation must not be made between figures on different sides of the lines.

For dry bulb temperatures below 0°C (32°F) it will be noticed that, when the depression of the wet bulb is zero, i.e. when the temperature of the wet bulb is equal to that of the dry bulb, the dew-point is still below the dry bulb, and the relative humidity is less than 100 per cent. These apparent anomalies are a consequence of the method of computing dew-points and relative humidities now adopted by the Meteorological Office, in which the standard saturation pressure for temperatures below 0°C (32°F) is taken as that over water, and not as that over ice.

TABLE 16

DEW-POINT (°C)

(FOR USE WITH ASPIRATED PSYCHROMETER)

Dry Bulb °C	Depression of Wet Bulb																			Dry Bulb °C
	0°	0·5°	1·0°	1·5°	2·0°	2·5°	3·0°	3·5°	4·0°	4·5°	5·0°	5·5°	6·0°	6·5°	7·0°	7·5°	8·0°	8·5°	9·0°	
40	40	39	39	38	38	37	36	36	35	35	34	33	33	32	31	31	30	30	29	40
39	39	38	38	37	37	36	35	35	34	33	33	32	32	31	30	29	28	28	27	39
38	38	37	37	36	36	35	34	34	33	32	32	31	30	30	29	28	27	27	26	38
37	37	36	36	35	35	34	33	33	32	31	31	30	29	29	28	27	27	26	25	37
36	36	35	35	34	34	33	32	32	31	30	30	29	28	28	27	26	25	25	24	36
35	35	34	34	33	33	32	31	31	30	29	29	28	27	26	25	25	24	24	23	35
34	34	33	33	32	32	31	30	30	29	28	28	27	26	25	24	24	23	22	22	34
33	33	32	32	31	31	30	29	29	28	27	26	25	25	24	23	23	22	21	20	33
32	32	31	31	30	30	29	28	28	27	26	25	24	24	23	22	22	21	20	19	32
31	31	30	30	29	29	28	27	27	26	25	24	24	23	22	21	20	20	19	18	31
30	30	29	29	28	28	27	26	26	25	24	23	22	22	21	20	19	18	17	17	30
29	29	28	28	27	26	26	25	24	24	23	22	21	20	20	19	18	17	16	15	29
28	28	27	27	26	25	25	24	23	22	22	21	20	19	19	18	17	16	15	14	28
27	27	26	26	25	24	24	23	22	21	21	20	19	18	17	16	16	15	14	13	27
26	26	25	25	24	23	23	22	21	20	19	19	18	17	16	15	14	13	12	11	26
25	25	24	24	23	22	21	21	20	19	18	18	17	16	15	14	13	12	11	10	25
24	24	23	23	22	21	20	19	19	18	17	16	16	15	14	13	12	11	10	8	24
23	23	22	22	21	20	19	18	18	17	16	15	14	13	12	11	10	9	8	7	23
22	22	21	21	20	19	18	17	17	16	15	14	13	12	11	10	9	8	7	5	22
21	21	20	20	19	18	17	16	16	15	14	13	12	11	10	9	8	7	5	4	21
20	20	19	19	18	17	16	15	15	14	13	12	11	10	9	7	6	5	4	2	20
19	19	18	17	17	16	15	14	13	12	11	10	9	8	7	6	5	3	2	0	19
18	18	17	16	16	15	14	13	12	11	10	9	8	7	6	5	3	2	0	-1	18
17	17	16	15	15	14	13	12	11	10	9	8	7	6	5	4	2	0	-2	-3	17
16	16	15	14	14	13	12	11	10	9	8	7	5	4	3	2	0	-2	-4	-6	16
15	15	14	13	13	12	11	10	9	8	7	5	4	3	1	0	-2	-4	-6	-8	15
14	14	13	12	12	11	10	9	8	7	5	4	3	1	0	-2	-4	-6	-8	-11	14
13	13	12	11	11	10	9	8	7	6	4	3	1	0	-2	-4	-6	-8	-11	-14	13
12	12	11	10	10	9	8	7	6	5	3	1	0	-2	-4	-6	-8	-10	-13	-17	12
11	11	10	9	9	8	7	6	5	4	2	0	-2	-3	-5	-6	-10	-13	-17	-22	11
10	10	9	8	8	7	6	5	4	3	1	-2	-3	-5	-7	-10	-13	-16	-21	-29	10

140

TABLE 16—(contd)

Depression of Wet Bulb

Dry Bulb °C	0°	0·5°	1·0°	1·5°	2·0°	2·5°	3·0°	3·5°	4·0°	4·5°	5·0°	5·5°	6·0°	6·5°	7·0°	7·5°	8·0°	8·5°	9·0°	Dry Bulb °C
9	9	8	7	6	5	4	3	1	0	−2	−3	−5	−7	−9	−12	−16	−20	−27	−45	9
8	8	7	6	5	4	3	1	0	−2	−3	−5	−7	−9	−12	−15	−19	−25	−25	−36	8
7	7	6	5	4	3	1	0	−1	−3	−5	−7	−9	−11	−14	−18	−19	−24	−34		7
6	6	5	4	3	1	0	−1	−3	−4	−6	−8	−11	−14	−14	−18	−23	−32			6
5	5	4	3	2	0	−1	−3	−4	−6	−8	−10	−11	−14	−18	−22	−30				5
4	4	3	2	0	−1	−2	−4	−6	−8	−9	−11	−14	−17	−22	−28	−45				4
3	3	2	1	−1	−2	−4	−5	−6	−8	−11	−13	−16	−21	−27	−39					3
2	2	1	0	−2	−3	−4	−6	−8	−10	−13	−16	−20	−25	−34						2
1	1	0	−2	−3	−4	−6	−8	−10	−12	−15	−19	−24	−31							1
0	0	−1	−3	−4	−6	−8	−9	−12	−14	−18	−22	−29	−44							0
−1	−1	−2	−4	−5	−7	−9	−11	−14	−17	−21	−26	−37								
−2	−2	−4	−5	−7	−9	−11	−13	−16	−19	−24	−32									
−3	−3	−5	−6	−8	−10	−12	−15	−18	−23	−29	−44									
−4	−4	−7	−9	−10	−12	−14	−17	−21	−26	−36										
−5	−5	−8	−10	−11	−13	−16	−19	−24	−31											
−6	−6	−10	−12	−13	−15	−18	−22	−28	−39											
−7	−7	−11	−13	−14	−17	−20	−25	−32												
−8	−8	−12	−14	−16	−19	−23	−28	−40												
−9	−9	−13	−16	−17	−21	−25	−33													
−10	−10	−15	−17	−19	−23	−28	−39													
−11	−11	−16	−19	−21	−25	−32														
−12	−12	−17	−20	−23	−28	−38														
−13	−13	−18	−22	−25	−31	−47														
−14	−14	−20	−24	−27	−35															
−15	−15	−21	−25	−29	−40															
−16	−16	−22	−27	−32																
−17	−17			−35																

See footnotes to Table 15 (page 139)

TABLE 17

DEW-POINT (°F)

(FOR USE WITH MARINE SCREEN)

Dry Bulb °F	Depression of Wet Bulb																						Dry Bulb °F
	0	0·5	1	1·5	2	2·5	3	3·5	4	4·5	5	6	7	8	9	10	11	12	13	14	15	16	
100	100	99	99	99	98	97	96	95	95	95	94	92	91	90	88	87	86	84	83	81	80	78	100
99	99	99	98	97	97	96	95	95	94	93	93	91	90	89	87	86	84	83	81	80	78	77	99
98	98	97	97	96	95	95	94	94	93	92	92	90	89	88	86	85	83	82	80	79	77	76	98
97	97	96	96	95	94	94	93	93	92	91	91	89	88	87	85	84	82	81	79	78	76	74	97
96	96	95	95	94	94	93	92	92	91	90	90	88	87	85	84	83	81	80	78	76	75	73	96
95	95	94	94	93	92	92	91	90	90	89	89	87	86	84	83	81	80	78	77	75	74	72	95
94	94	93	93	92	91	91	90	89	89	88	87	86	85	83	82	80	79	77	76	74	72	71	94
93	93	92	92	91	90	90	89	88	88	87	86	85	84	82	81	79	78	76	75	73	71	69	93
92	92	91	91	90	89	89	88	87	87	86	85	84	83	81	80	78	77	75	73	72	70	68	92
91	91	90	90	89	88	88	87	86	86	85	84	83	81	80	79	77	75	74	72	70	69	67	91
90	90	89	89	88	87	87	86	85	85	84	83	82	80	79	77	76	74	73	71	69	68	66	90
89	89	88	88	87	86	85	85	84	84	83	82	81	79	78	76	75	73	71	70	68	66	64	89
88	88	87	87	86	85	85	84	83	83	82	81	80	78	77	75	74	72	70	69	67	65	63	88
87	87	86	86	85	84	84	83	82	81	81	80	79	77	76	74	72	71	69	67	66	64	62	87
86	86	85	85	84	83	83	82	81	80	80	79	78	76	75	73	71	70	68	66	64	62	60	86
85	85	84	84	83	82	82	81	80	79	79	78	77	75	73	72	70	69	67	65	63	61	59	85
84	84	83	83	82	81	81	80	79	78	78	77	75	74	72	71	69	67	66	64	62	60	58	84
83	83	82	82	81	80	79	79	78	77	77	76	74	73	71	70	68	66	64	62	60	58	56	83
82	82	81	81	80	79	79	78	77	76	76	75	73	72	70	68	67	65	63	61	59	57	55	82
81	81	80	80	79	78	77	77	76	75	75	74	72	71	69	67	66	64	62	60	57	55	53	81
80	80	79	79	78	77	76	76	75	74	74	73	71	69	68	66	64	62	60	59	56	54	52	80
79	79	78	78	77	76	75	75	74	73	72	72	70	68	67	65	63	61	59	57	55	53	50	79
78	78	77	77	76	75	74	74	73	72	71	71	69	67	66	64	62	60	58	56	54	52	49	78
77	77	76	76	75	74	73	73	72	71	70	69	68	66	64	63	61	59	57	55	52	50	48	77
76	76	75	75	74	73	72	72	71	70	69	68	67	65	63	61	60	57	55	53	51	48	46	76
75	75	74	74	73	72	71	71	70	69	68	67	66	64	62	60	58	56	54	52	50	47	44	75
74	74	73	73	72	71	70	69	69	68	67	66	64	63	61	59	57	55	53	50	48	45	43	74
73	73	72	72	71	70	69	68	68	67	66	65	63	62	60	58	56	54	52	49	47	44	41	73
72	72	71	70	70	69	68	67	66	66	65	64	62	61	59	57	55	52	50	48	45	42	39	72
71	71	70	69	69	68	67	67	65	65	64	63	61	59	57	55	53	51	49	46	43	41	38	71
70	70	69	68	68	67	66	65	64	63	63	62	60	58	56	54	52	50	47	45	42	39	36	70
69	69	68	67	67	66	65	64	63	62	61	61	59	57	55	53	51	48	46	43	40	38	34	69
68	68	67	66	66	65	64	63	62	61	61	60	58	56	54	52	49	47	44	42	39	36	32	68
67	67	66	65	65	64	63	62	61	60	59	59	57	55	53	50	48	46	43	40	37	34	30	67
66	66	65	64	64	63	62	61	60	59	58	57	55	53	51	49	47	44	41	38	35	32	28	66
65	65	64	63	63	62	61	60	59	58	57	56	54	52	50	48	45	43	40	37	34	30	26	65
64	64	63	62	62	61	60	59	58	57	56	55	53	51	49	47	44	41	38	35	32	28	23	64
63	63	62	61	61	60	59	58	57	56	55	54	52	50	48	45	43	40	37	34	30	26	21	63
62	62	61	60	59	59	58	57	56	55	54	53	51	49	46	44	41	38	35	31	28	23	18	62
61	61	60	59	58	58	57	56	55	54	53	52	50	47	45	42	40	37	34	30	26	21	16	61
60	60	59	58	57	56	55	54	53	52	51	50	48	46	44	41	38	35	32	28	23	19	13	60
59	59	58	57	56	55	54	53	52	51	50	49	47	45	42	40	37	34	30	26	22	16	9	59
58	58	57	56	55	54	53	52	51	50	49	48	46	43	41	38	35	32	28	24	19	13	5	58
57	57	56	55	54	53	52	51	50	49	48	47	45	42	40	37	34	30	26	22	16	10	1	57
56	56	55	54	53	52	51	50	49	48	47	46	43	41	38	35	32	29	25	20	14	7		56
55	55	54	53	52	51	50	49	48	47	46	45	42	40	37	34	30	26	22	17	11	3		55
54	54	53	52	51	50	49	48	47	46	45	43	41	38	35	32	29	25	20	14	8			54
53	53	52	51	50	49	48	47	46	45	43	42	40	37	34	31	27	23	18	12	4			53
52	52	51	50	49	48	47	46	45	43	42	41	38	35	32	29	25	20	16	9	0			52
51	51	50	49	48	47	46	45	43	42	41	40	37	34	31	27	23	19	13	6				51
50	50	49	48	47	46	45	43	42	41	40	39	36	33	29	25	21	16	10	2				50
49	49	48	47	46	45	44	42	41	40	39	38	35	32	28	24	19	14	7					49
48	48	47	46	45	44	43	41	40	39	38	36	33	30	26	22	17	12	4					48
47	47	46	45	44	43	42	40	39	38	36	35	32	28	24	20	15	9	0					47
46	46	45	44	43	42	40	39	38	36	35	34	30	27	23	19	13	6						46
45	45	44	43	42	40	39	38	36	35	34	32	29	25	21	16	10	3						45
44	44	43	42	41	39	38	37	35	34	32	31	28	23	19	15	8							44
43	43	42	41	39	38	37	36	34	33	31	29	26	22	17	12	4							43
42	42	41	40	38	37	36	34	33	32	30	28	25	20	16	9	3							42
41	41	40	39	37	36	35	33	32	30	29	27	23	18	13	7	$\overline{4}$							41
40	40	39	38	36	35	34	32	31	29	27	26	22	17	$\underline{11}$	$\overline{8}$	1							40

142

TABLE 17—(contd)

Dry Bulb °F	Depression of Wet Bulb																						Dry Bulb °F
°F	0	0·5	1	1·5	2	2·5	3	3·5	4	4·5	5	6	7	8	9	10	11	12	13	14	15	16	°F
39	39	38	36	35	34	32	31	29	28	26	24	20	15	12	6								39
38	38	37	35	34	33	31	30	28	26	24	22	18	15	10	3								38
37	37	36	34	33	32	30	29	27	25	23	21	18	13	7									37
36	36	35	33	32	30	29	27	25	23	23	21	16	11	5									36
35	35	34	32	31	29	28	26	25	23	21	19	14	9	3									35
34	34	33	31	30	28	27	25	23	22	20	17	13	7										34
33	33	32	30	29	27	26	24	22	20	18	16	11	4										33
32	32	31	29	28	26	24	22	21	19	16	14	8	1										32
31	31	29	28	27	25	23	21	19	17	14	12	6											31
30	30	29	27	25	23	22	20	17	15	13	10	4											30
29	29	27	26	24	22	20	18	16	14	11	8	1											29
28	28	26	24	22	20	18	16	14	12	9	6												28
27	27	25	23	21	19	17	15	13	10	7	4												27
26	25	23	22	20	18	16	14	11	8	5	2												26
25	24	22	21	19	17	14	12	10	7	3													25
24	23	21	19	17	15	13	11	8	5	1													24
23	22	20	18	16	14	12	9	6	3														23
22	21	19	17	15	13	10	7	4	1														22
21	20	18	16	13	11	9	6	2															21
20	19	16	14	12	10	7	4	0															20
19	17	15	13	11	8	5	2																19
18	16	14	12	10	7	4	0																18
17	15	13	11	8	5	2																	17
16	14	12	10	7	4																		16
15	13	11	8	5	2																		15
14	12	10	7	4	1																		14
13	11	8	6	3																			13
12	10	7	4	1																			12
11	9	6	3																				11
10	7	5	2																				10
9	6	3	0																				9
8	5	2																					8
7	4	1																					7
6	3	0																					6
5	2																						5
4	1																						4

See footnotes to Table 15 (page 139).

TABLE 18

DEW-POINT (°F)

(FOR USE WITH ASPIRATED PSYCHROMETER)

Dry Bulb °F	Depression of Wet Bulb																
	0°	1°	2°	3°	4°	5°	6°	7°	8°	9°	10°	11°	12°	13°	14°	15°	16°
100	100	99	98	96	95	94	93	91	90	89	87	86	85	83	82	81	79
98	98	97	96	94	93	92	91	89	88	87	85	84	83	81	80	78	77
96	96	95	94	92	91	90	89	87	86	85	83	82	80	79	77	76	74
94	94	93	92	90	89	88	86	85	84	82	81	80	78	77	75	74	72
92	92	91	89	88	87	86	84	83	82	80	79	77	76	74	73	71	70
90	90	89	87	86	85	84	82	81	79	78	77	75	74	72	71	69	67
88	88	87	85	84	83	81	80	79	77	76	74	73	71	70	68	67	65
86	86	85	83	82	81	79	78	77	75	74	72	71	69	67	66	64	62
84	84	83	81	80	79	77	76	74	73	71	70	68	67	65	63	61	60
82	82	81	79	78	77	75	74	72	71	69	68	66	64	63	61	59	57
80	80	79	77	76	75	73	72	70	69	67	65	64	62	60	58	56	54
78	78	77	75	74	72	71	69	68	66	65	63	61	60	58	56	54	52
76	76	75	73	72	70	69	67	66	64	62	61	59	57	55	53	51	49
74	74	73	71	70	68	67	65	64	62	60	58	57	55	53	50	48	46
72	72	71	69	68	66	65	63	61	60	58	56	54	52	50	48	45	43
70	70	69	67	66	64	62	61	59	57	55	54	51	49	47	45	42	40
68	68	67	65	63	62	60	59	57	55	53	51	49	47	44	42	39	36
66	66	64	63	61	60	58	56	54	53	51	49	46	44	42	39	36	33
64	64	62	61	59	58	56	54	52	50	48	46	44	41	39	36	33	29
62	62	60	59	57	55	54	52	50	48	46	43	41	38	35	32	29	25
60	60	58	57	55	53	51	49	47	45	43	41	38	35	32	29	25	21
58	58	56	55	53	51	49	47	45	43	40	38	35	32	29	25	21	16
56	56	54	53	51	49	47	45	42	40	38	35	32	29	25	21	16	11
54	54	52	50	49	47	45	42	40	37	35	32	29	25	21	17	11	11
52	52	50	48	46	44	42	40	37	35	32	29	25	21	17	11	5	4

144

TABLE 18—(contd)

Depression of Wet Bulb

Dry Bulb °F	0°	1°	2°	3°	4°	5°	6°	7°	8°	9°	10°	11°	12°	13°	14°	15°	16°
50	50	48	46	44	42	40	37	35	32	29	26	22	17	12	6		
48	48	46	44	42	40	37	35	32	29	26	22	18	13	7			
46	46	44	42	40	37	35	32	29	26	22	18	14	8	1			
44	44	42	40	37	35	32	30	26	23	19	14	9	2	0			
42	42	40	38	35	33	30	27	23	19	15	10	8	2				
40	40	38	35	33	30	27	24	20	16	14	9	2					
38	38	36	33	31	28	25	21	19	15	9	4						
36	36	34	31	28	25	23	19	15	10	5							
34	34	32	29	26	23	20	16	11	6								
32	32	29	27	23	20	17	12	7	1								

Depression of Wet Bulb

Dry Bulb °F	0°	0·5°	1·0°	1·5°	2·0°	2·5°	3·0°	3·5°	4·0°	4·5°	5·0°	5·5°	6·0°	6·5°	7·0°	7·5°
30	30	29	27	26	24	23	21	19	17	15	13	11	8	5	3	0
28	28	26	25	23	22	20	18	16	14	12	10	7	4	1		
26	25	24	22	21	19	17	15	13	11	9	6	4	0			
24	23	21	20	18	16	14	12	10	8	5	2					
22	21	19	17	16	14	11	9	7	4	1						
20	19	17	15	13	11	9	6	4	1							
18	16	14	12	10	8	6	3	0								
16	14	12	10	8	5	3	0									
14	12	10	7	5	2	0										
12	10	7	5	2	0											

See footnotes to Table 15 (page 139)

145

TABLE 19

RELATIVE HUMIDITY (per cent)
(FOR USE IN A MARINE SCREEN)

Depression of Wet Bulb (degrees Celsius)

Dry Bulb °C	0·2°	0·4°	0·6°	0·8°	1·0°	1·2°	1·4°	1·6°	1·8°	2·0°	2·5°	3·0°	3·5°	4·0°	4·5°	5·0°	5·5°	6·0°	6·5°	7·0°	7·5°	8·0°	8·5°	9·0°	Dry Bulb °C
40	99	97	96	95	94	92	91	90	89	88	85	82	79	76	73	71	68	66	63	61	58	56	53	51	40
39	99	97	96	95	94	92	91	90	89	87	84	82	79	76	73	70	68	65	63	60	58	55	53	50	39
38	99	97	96	95	94	92	91	90	89	87	84	81	78	75	73	70	67	65	62	59	57	54	52	50	38
37	99	97	96	95	93	92	91	90	88	87	84	81	78	75	72	69	67	64	61	59	56	54	51	49	37
36	99	97	96	95	93	92	91	89	88	87	84	80	77	74	72	69	66	63	61	58	55	53	50	48	36
35	99	97	96	95	93	92	91	89	88	87	83	80	77	74	71	68	65	63	60	57	55	52	49	47	35
34	99	97	96	94	93	91	91	89	88	86	83	80	76	73	70	68	64	62	59	56	53	50	48	46	34
33	99	97	96	94	93	91	90	89	87	86	83	80	76	73	70	67	64	61	58	55	53	50	47	45	33
32	99	97	96	94	93	91	90	88	87	86	82	79	75	72	69	66	63	60	57	54	52	48	46	44	32
31	99	97	96	93	93	91	90	88	87	86	82	79	75	72	69	66	63	60	57	54	51	47	44	43	31
30	98	97	95	93	93	91	90	88	87	85	82	78	75	72	68	65	62	59	56	53	50	47	44	42	30
29	98	97	95	94	92	91	89	88	86	85	81	78	74	71	68	65	61	58	55	52	49	46	43	40	29
28	98	97	95	94	92	91	89	88	86	85	81	77	74	70	67	64	60	57	53	50	48	45	42	39	28
27	98	97	95	94	92	90	89	87	86	84	81	77	73	70	66	63	59	55	52	49	47	44	41	38	27
26	98	97	95	93	92	90	89	87	85	84	80	76	73	69	66	62	59	55	51	47	46	42	39	36	26
25	98	97	95	93	92	90	88	87	85	84	80	76	72	68	65	61	57	54	50	46	44	41	38	35	25
24	98	97	95	93	91	90	88	86	85	83	79	75	71	68	64	60	57	53	48	45	43	39	36	33	24
23	98	96	95	93	91	89	88	86	84	83	79	75	71	67	63	59	56	52	47	43	41	38	35	31	23
22	98	96	95	93	91	89	88	85	84	82	78	74	70	66	62	58	54	51	45	42	40	34	31	29	22
21	98	97	94	93	91	89	87	85	84	82	77	73	69	65	61	57	53	49	45	41	38	34	31	27	21
20	97	96	94	92	91	89	87	85	83	81	77	73	68	64	60	56	52	48	44	40	36	33	29	25	20
19	98	96	94	92	90	88	86	85	83	81	76	72	67	63	59	55	50	46	42	38	34	31	27	23	19
18	98	96	94	92	90	88	86	84	82	80	76	71	66	62	58	53	49	45	41	36	32	28	25	21	18
17	98	96	94	90	90	88	86	84	82	80	75	70	65	61	56	52	47	43	39	34	30	26	22	18	17
16	98	96	93	89	89	87	85	83	81	79	74	69	64	60	55	50	46	41	37	32	28	24	17	13	16
15	98	95	93	89	89	86	84	83	80	78	73	68	63	58	53	49	44	39	35	30	26	21	14	10	15
14	98	95	93	88	89	86	84	82	80	77	72	67	62	57	52	47	42	37	32	28	23	18	11	6	14
13	98	95	93	88	88	86	83	81	79	77	71	66	61	55	50	45	40	35	30	25	20	16	8	3	13
12	97	95	92	87	88	85	83	80	78	76	70	65	59	54	48	43	38	32	27	22	17	12			12
11	97	95	92	87	87	85	82	80	78	75	69	63	58	52	46	41	35	30	25	19	14	9			11
10	97	95	92	87	87	84	82	79	77	74	68	62	56	50	44	38	33	27	22	16	11	5			10

146

TABLE 19—(contd)

RELATIVE HUMIDITY (per cent)

Dry Bulb °C	Depression of Wet Bulb (degrees Celsius)																					
	0·2°	0·4°	0·6°	0·8°	1·0°	1·2°	1·4°	1·6°	1·8°	2·0°	2·5°	3·0°	3·5°	4·0°	4·5°	5·0°	5·5°	6·0°	6·5°	7·0°	7·5°	8·0°
9	97	95	92	89	86	84	81	79	76	73	67	61	54	48	42	36	30	24	18	13	7	2
8	97	94	92	89	86	83	80	78	75	72	66	59	52	46	40	33	27	21	15	9	3	
7	97	94	91	88	85	82	80	77	74	71	64	57	50	44	37	31	24	18	11	5	5	
6	97	94	91	88	85	82	79	76	73	70	63	55	48	41	34	28	21	14	13	6		
5	97	94	90	87	84	81	78	75	72	69	61	53	46	39	31	24	22	15	8	2		
4	97	93	90	86	83	80	77	74	70	67	59	51	44	36	32	25	18	10	3			
3	96	93	89	86	83	79	76	72	69	66	57	49	44	36	28	21	13	6				
2	96	93	89	85	82	78	75	71	67	64	57	49	41	33	24	16	9	1				
1	96	92	88	85	81	78	75	71	67	64	55	46	37	29	20	12	3					
0	96	92	88	84	80	76	73	69	65	61	52	43	33	24	16	7						
−1	95	91	87	83	78	74	70	66	62	58	49	39	29	20	11	1						
−2	94	89	85	81	77	72	68	64	60	56	45	35	25	15	5							
−3	93	88	84	79	75	70	66	61	57	53	42	31	21	10								
−4	91	87	82	77	72	68	63	59	54	49	38	27	16	5								
−5	90	85	80	75	70	65	61	56	51	46	34	22	11									
−6	89	84	79	73	68	63	58	53	48	42	30	17	5									
−7	88	82	77	71	66	60	55	49	44	39	25	12										
−8	87	81	75	69	63	57	52	46	40	35	20	6										
−9	85	79	73	67	61	54	48	42	36	30	15											
−10	84	77	71	64	58	51	45	38	32	26												
−11	83	76	69	62	55	48	41	34	27	21												
−12	81	74	67	59	52	44	37	30	22	15												
−13	80	72	64	56	48	41	33	25	17	9												
−14	79	70	62	53	45	36	28	20	11	3												
−15	77	68	59	50	41	32	23	14	5													
−16	76	66	56	47	37	27	18	8	2													
−17	74	64	53	43	33	22	12	2														

In the tables, lines are ruled to draw attention to the fact that above the line evaporation is going on from a water surface, while below the line it is going on from an ice surface (wet-bulb temperature below 0°C). Owing to this, interpolation must not be made between figures on different sides of the line.

TABLE 20

CONVERSION OF TEMPERATURE READINGS ON THE FAHRENHEIT SCALE TO THE CELSIUS
(FORMERLY 'CENTIGRADE') AND KELVIN (FORMERLY 'ABSOLUTE') SCALES

°F	°C	K	°F	°C	K	°F	°C	K
0	−17·8	255·35	40	+ 4·4	277·55	80	+26·7	299·85
1	17·2	55·95	41	5·0	78·15	81	27·2	300·35
2	16·7	56·45	42	5·6	78·75	82	27·8	0·95
3	16·1	57·05	43	6·1	79·25	83	28·3	1·45
4	15·6	57·55	44	6·7	79·85	84	28·9	2·05
5	15·0	58·15	45	7·2	80·35	85	29·4	2·55
6	14·4	58·75	46	7·8	80·95	86	30·0	3·15
7	13·9	59·25	47	8·3	81·45	87	30·6	3·75
8	13·3	59·85	48	8·9	82·05	88	31·1	4·25
9	12·8	260·35	49	9·4	282·55	89	31·7	304·85
10	12·2	260·95	50	10·0	283·15	90	32·2	305·35
11	11·7	61·45	51	10·6	83·75	91	32·8	5·95
12	11·1	62·05	52	11·1	84·25	92	33·3	6·45
13	10·6	62·55	53	11·7	84·85	93	33·9	7·05
14	10·0	63·15	54	12·2	85·35	94	34·4	7·55
15	9·4	63·75	55	12·8	85·95	95	35·0	8·15
16	8·9	64·25	56	13·3	86·45	96	35·6	8·75
17	8·3	64·85	57	13·9	87·05	97	36·1	9·25
18	7·8	65·35	58	14·4	87·55	98	36·7	9·85
19	7·2	265·95	59	15·0	288·15	99	37·2	310·35
20	6·7	266·45	60	15·6	288·75	100	37·8	310·95
21	6·1	67·05	61	16·1	89·25	101	38·3	11·45
22	5·6	67·55	62	16·7	89·85	102	38·9	12·05
23	5·0	68·15	63	17·2	90·35	103	39·4	12·55
24	4·4	68·75	64	17·8	90·95	104	40·0	13·15
25	3·9	69·25	65	18·3	91·45	105	40·6	13·75
26	3·3	69·85	66	18·9	92·05	106	41·1	14·25
27	2·8	70·35	67	19·4	92·55	107	41·7	14·85
28	2·2	70·95	68	20·0	93·15	108	42·2	15·35
29	1·7	271·45	69	20·6	293·75	109	42·8	315·95
30	1·1	272·05	70	21·1	294·25	110	43·3	316·45
31	−0·6	72·55	71	21·7	94·85	111	43·9	17·05
32	0·0	73·15	72	22·2	95·35	112	44·4	17·55
33	+0·6	73·75	73	22·8	95·95	113	45·0	18·15
34	1·1	74·25	74	23·3	96·45	114	45·6	18·75
35	1·7	74·85	75	23·9	97·05	115	46·1	19·25
36	2·2	75·35	76	24·4	97·55	116	46·7	19·85
37	2·8	75·95	77	25·0	98·15	117	47·2	20·35
38	3·3	76·45	78	25·6	98·75	118	47·8	20·95
39	+3·9	277·05	79	+26·1	299·25	119	+48·3	321·45

TABLE 21

CONVERSION OF NAUTICAL MILES TO KILOMETRES

Nautical Miles	Kilometres	Nautical Miles	Kilometres
1	1·9	20	37
2	3·7	30	56
3	5·6	40	74
4	7·4	50	93
5	9·3	60	111
6	11·1	70	130
7	13·0	80	148
8	14·8	90	167
9	16·7	100	185
10	18·5		

Based on the International Nautical Mile of 1852 m.

TABLE 22

CONVERSION OF FEET TO METRES

Feet	Metres	Feet	Metres	Feet	Metres	Feet	Metres
1	0·30	20	6·1	200	61	2 000	610
2	0·61	30	9·1	300	91	3 000	910
3	0·91	40	12·2	400	122	4 000	1 220
4	1·22	50	15·2	500	152	5 000	1 520
5	1·52	60	18·3	600	183	6 000	1 830
6	1·83	70	21·3	700	213	7 000	2 130
7	2·13	80	24·4	800	244	8 000	2 440
8	2·44	90	27·4	900	274	9 000	2 740
9	2·74	100	30·5	1 000	305	10 000	3 050
10	3·05						

APPENDIX

THE INTERNATIONAL SYSTEM OF UNITS
(Système International d'Unités)

The International System (SI) consists of seven 'base units' together with two 'supplementary units'. From these are formed others known as 'derived units'. The base and supplementary units, and some of the derived units, have been given names and symbols. The symbols are printed in lower case except where derived from the name of a person; for example m (metre), but A (ampere). Symbols are not pluralized (1 m, 10 m) nor do they take a full stop. The names of the units do not, however, take capitals (except of course at the beginning of a sentence), although they may be pluralized; for example, 1 kelvin, 10 kelvins.

The base units are:

metre (symbol m)	the unit of length
kilogram (symbol kg)	the unit of mass
second (symbol s)	the unit of time
ampere (symbol A)	the unit of electrical current
kelvin (symbol K)	the unit of thermodynamic temperature, defined as the fraction $1/273 \cdot 16$ of the thermodymanic temperature of the triple point of water.
candela (symbol cd)	the unit of luminous intensity
mole (symbol mol)	the unit of the amount of a substance which contains the same number of molecules as there are atoms in exactly 12 grams of pure carbon.

The two supplementary units are:

radian (symbol rad)	the measure of a plane angle
steradian (symbol sr)	the measure of a solid angle

A few of the derived units are:

Quantity	Name of unit	Symbol	Expressed in base units
frequency	hertz	Hz	$1 \text{ Hz} = 1 \text{ s}^{-1}$
force	newton	N	$1 \text{ N} = 1 \text{ kg m/s}^2$
pressure	pascal	Pa	$1 \text{ Pa} = 1 \text{ N/m}^2$
work	joule	J	$1 \text{ J} = 1 \text{ N m}$
power	watt	W	$1 \text{ W} = 1 \text{ J/s}$

(1 newton = 10^5 dynes, 1 pascal = 10^{-2} millibars, 1 joule = 10^7 ergs).

NON–SI UNITS

The following non-SI units are in current use in the Meteorological Office and may be found in publications of the Office.

1. Pressure

The millibar is used as the unit of pressure in meteorology. Despite the recommended abbreviation mbar, the Meteorological Office will continue to use mb, (1 mb=1 hPa, where h=hecto=10^2).

2. Temperature

The unit Celsius (symbol °C) continues to be used.

Celsius temperature = temperature (in kelvins) minus 273·15 K (note that the sign ° is no longer used with K).

A difference in temperature should be expressed by use of the international symbol 'deg' without a qualifying C or K.

3. Distance

There is a continuing requirement for some distances to be measured in nautical miles (symbol n. mile).

Because the nautical mile varies with latitude, an internationally agreed International Nautical Mile is preferred. This has been in use in the United Kingdom since 1970.

The International Nautical Mile is defined as 1852 m (6076·12 feet).

4. Height

Heights other than cloud heights are expressed in metres. Because of the requirements of aviation the heights of cloud will continue for the time being to be expressed in feet (1 foot = 0·3048 m).

5. Speed

The derived SI unit is the metre per second (m/s). However, the World Meteorological Organization recommends the use of the knot for horizontal wind speed for the time being (1 knot = 1 nautical mile per hour = 0·5 m/s). The symbol kn for knot is recommended to avoid confusion with the symbol for kilotonne and will be used in Meteorological Office publications.

6. Time

Units other than SI, such as day, week, month and year, are in common use.

7. Direction

Direction is measured in degrees clockwise from north and refers to the true compass, indicated by the symbol °T, for example 320°T.

8. Cloud amounts

The use of 'okta' for the measurement of cloud amount is authorized by the World Meteorological Organization.

INDEX

153

INDEX—*contd.*